Neil Kaminar

SOLAR BASICS

THE EASY GUIDE TO SOLAR ENERGY

website: www.thesolardesignbook.com
email: postmaster@thesolardesignbook.com

Copyright © 2009 McNeill Hill Publications

ALL RIGHTS RESERVED. No part of this book may be reproduced or transmitted in any form by any means, electronic or mechanical, including photocopying and recording, or by any information storage and retrieval system, except as may be expressly permitted by the 1976 Copyright Act or by the publisher.

Cover photographs courtesy the US Department of Energy and the National Renewable Energy Laboratory (DOE/NREL). These photographs and others in the book from DOE/NREL do not imply affiliation with DOE/NREL. Front cover credit: Solar Design Associates and rear cover credit: William Lord. Unless otherwise noted, photographs and illustrations are by the author.

Disclaimer

This book is not intended to endorse any product or technology or make decisions for the reader. Nor is it intended to make an expert of the reader. It does not guarantee that the reader will have a successful experience sizing, buying, installing, or servicing any solar system. Nor does the book guarantee that the reader will be able to pick the right installer. Grid-connected systems require a licensed installer, permits, adherence to codes, inspection, and permission by the local utility to hook up to the grid. Electrical systems can be dangerous. The author or publisher is not responsible for any losses incurred as a result of acting on the advice of this book.

Publisher's Cataloging-in-Publication data

Kaminar, Neil Randel.
    Solar basics / Neil Kaminar.
    p. cm.
    Includes bibliographical references and index.
    ISBN 978-0-9840510-0-7
1. Solar energy. 2. Solar houses. 3. Solar cells. 4. Photovoltaic energy sources. 5. Solar pumps. 6. Solar energy--Equipment and supplies. I. Title.
TJ810 .K36 2009
621.47—dc22   2009934423

Printed in the United States of America by BookPrintingRevolution.com

Acknowledgements

Many thanks to all the people who have reviewed the early versions of the book. Special thanks to John and Christina for editing and comments. Special thanks to Lisa for helping with the cover design. And extra special thanks to Fran for the encouragement, editing, and understanding.

About *Solar Design*

The companion book, *Solar Design* is written for the solar professional or dedicated do-it-yourselfer. The book goes into detail about how to design a solar system, including sizing the solar array and choosing equipment.

The book includes a CD with software to size the solar array or predict the output of a planned solar array. There is a data base of over 270 sites in the U.S. states and territories. The software allows arrays to be fixed at any orientation and tilt. It also allows any kind of tracking. There is even provision for predicting the output of concentrator modules.

Other valuable software is included, such as the Wiring Calculator that helps the designer size wires and breakers. Additional software allows the designer to generate a water pump curve, calculate module efficiency, do a costing analysis, and determine shadowing between fixed rows of panels.

The software is designed to run in Microsoft Excel on any computer. Numerous examples are used to illustrate the software and concepts in the book. There are chapters on how solar cells work and how modules are made.

Chapters

1. About The Software
2. Sizing a Solar System
3. Example 1, Solar Water System
4. Example 2, Grid-Connected Residence 1
5. Example 3, Battery System
6. Example 4, Grid-Connected Residence 2
7. Example 5, Sailboat
8. Example 6, Grid-Connected with Battery Backup
9. How Solar Cells Work
10. How Modules are Made

For more information go to **www.TheSolarDesignBook.com**

TABLE OF CONTENTS

About Solar Design iv
Table of Contents v
List of Illustrations xi
List of Tables xv

Foreword 1

Introduction

    Why Solar 3
    The Purpose of This Book 4
    Terminology 4
    Price and Model Numbers 5

Chapter 1   Solar Module Basics

    What is a Solar Module 7
    Solar Modules Use Sunlight 8
    What Comes Out of the Wires 14
    Can You Harm Them 16

Chapter 2   Applications

    Two Basic Types of Applications 19
    Grid-Connected Systems 19
    Stand-Alone Battery Systems 22
    Grid-Connected with Battery Backup 26
    Solar Powered Water Systems 27
    Other Applications 29

Chapter 3   Types of Modules

    How Solar Modules Work 31
    Crystalline Silicon Modules 32
    Back Contact Solar Cells 34
    Thin-Film Solar Cells 35

Other Types of Solar Cells 38
Concentrator Modules 38
Module Efficiency 41

Chapter 4 Inverters

Two Basic Types of Inverters 43
Grid-Connected Inverters 43
Maximum Peak Power Tracking 45
Stand-Alone Inverters 46
Non Sine Wave Inverters 47
Low Voltage Disconnect 47
Hybrid Inverters 48
Surge Current 48
Inverter Efficiency 49
High Temperature Derating 50
Special Features 50
Stacking for 240 VAC 50
VA Capacity and Power Factor 51
Ground Fault Circuit Interrupters 51
Battery Charging Built Into Inverter 52
Display 52
Data Link 52
NEC Required Features 52
Certifications 52
How to Pick an Inverter 54

Chapter 5 Batteries

Battery Basics 55
Types of Rechargeable Batteries 56
Types of Lead-Acid Batteries 57
Starting Battery vs. Deep Discharge Battery 62
Wiring 63
Expense and Maintenance 64
Charging a Lead-Acid Battery 65
Battery Temperature Compensation 66

TABLE OF CONTENTS   VII

    Battery Desulfator 67
    Aging 67
    Additives 67
    Temperature 67
    How to Choose Batteries 68

Chapter 6   Charge Controllers

    The Function of a Charge Controller 69
    Types of Charge Controllers 70
    Pulse Width Modulation 72
    Diversion Charge Controllers 72
    Multiple Source Charge Controllers 73
    Charge Controller Ratings 73
    Temperature Derating 74
    Temperature Compensation 74
    Charging Voltage 74
    Low Voltage Disconnect 74
    High Voltage Disconnect 75
    Parallel Configuration 75
    Efficiency 76
    Ground Fault Protection Device 77
    Mounting 77
    Display 77
    Data Link 77
    Certifications 78
    How to Pick a Charge Controller 78

Chapter 7   Solar Water Systems

    Types of Solar Water Systems 79
    Terminology 80
    Sources of Water 82
    Types of Pumps 85
    Discharge Curves 92
    Control Box 94
    Water Treatment 94

Reverse Osmosis 96
How to Pick a Pump 97

Chapter 8   The Basics of System Sizing

Energy Audit 99
Determine Loads 101
Account for Losses 102
Battery Sizing and Charging 103
Choose Equipment 103
Site Survey 104
Choose the Modules 105
Wiring 105
Balance-of-system 106
Documentation 106

Chapter 9   Federal and State Incentives

DSIRE 107
Federal Incentives 107
State Incentives 108
Loans 109
Grants 112
Tax Breaks 112
Production Incentives 112
Bonds 114
What To Do If Your State Does Not Have An Incentive Program . 114
The Future 114

Chapter 10   Sources

The Internet 115
How Search Engines Work 115
Collecting Bookmarks 116
Energy.Sourceguides.com 117
GoSolarCalifornia 118
Solar magazines 119

TABLE OF CONTENTS IX

    Books That List Sources 120
    Solar Exhibits and Fairs 121
    Cooler Planet 121
    NABCEP 121
    SEIR 122
    Word of Mouth 122

Chapter 11   Calculating Cost

    Types of Cost Calculations 123
    Discounting 123
    System Cost 124
    Life-cycle Cost 125
    Cost of Electricity 127
    Payback Period 128
    Intangible Credits 129

Chapter 12   Installation

    Mounting Structure 131
    Thermal Issues 138
    Installing Batteries 138
    Mechanical Wear and Tear 139
    Corrosion 141
    UV Radiation Damage 144
    Animals and Insects 145
    Security 148
    Quality Workmanship 148

Chapter 13   Instrumentation and Testing

    The Need for Instrumentation 149
    The Basics 150
    Measuring the Amount of Sunlight 154
    Measuring Temperature 156
    Measuring the IV Curve 159
    Correcting to Standard Conditions 161

Keeping Track of Energy Production 162
Measuring Battery State of Charge 164
Weather Stations 166
Summary 166

Chapter 14   Maintenance and Repair

Common Problems 167
Keeping it Clean 169
Shadows 170
Electrical Connections 171
Batteries 173
Overheating 175
Finding a Failed Module 176
Removing and Replacing a Failed Module 179
Recycling 180

Chapter 15   Safety

Electrical Safety 181
Battery Safety 188
Working on Roofs or at Heights 191
Working Safely with Hand Tools 195

Glossary 197

Index 207

About the Author 215

About *Solar Design* 217

LIST OF ILLUSTRATIONS

Figure 1	Bell Labs Engineer Testing a Solar Panel in 1954	1
Figure 2	1973 Satellite Solar Cell Used on Skylab Space Station	2
Figure 3	Front and Back of a Typical Solar Module	7
Figure 4	Shadows Greatly Reduce Module Output	8
Figure 5	Shadow-Producing Equipment. Radar Dome on a Boat Arch Mounted Solar System	9
Figure 6	Ventilation Under Modules	9
Figure 7	Annual US Solar Radiation	10
Figure 8	January US Solar Radiation	11
Figure 9	July US Solar Radiation	12
Figure 10	Flat Plate, Fixed Tilt, Facing South	13
Figure 11	Suntracker Model 30 Tracker with Four Solar Modules Mounted	14
Figure 12	Typical Label on the Back of a Solar Module	15
Figure 13	Bypass Diode Use	16
Figure 14	Module Glass After Intentional Breakage, No Cells	18
Figure 15	2 KW Residential Solar Array	20
Figure 16	Grid-connected Residential System Schematic	21
Figure 17	Direct Connected Battery Charging for a Remote Location	22
Figure 18	Solar Powered Battery System Schematic	23
Figure 19	Familiar Solar Powered Emergency Call Box	24
Figure 20	11 KW / Solar-Diesel Hybrid Village Power	25
Figure 21	Solar-Diesel System, Coiba National Park, Panama	25
Figure 22	Off-Grid Solar Powered House	26
Figure 23	Solar Powered Water System with Storage Tank	27
Figure 24	Solar Powered Livestock Watering System	28
Figure 25	Village Solar Water Pumping	29
Figure 26	Solar Powered Watch	30
Figure 27	Current Flow From a Solar Cell	31
Figure 28	Single Crystal Silicon Modules	32
Figure 29	Polycrystalline Silicon Module	33
Figure 30	Polycrystalline Silicon Cells Showing Individual Crystals	33
Figure 31	Back Contact Solar Cells from SunPower Corporation	34
Figure 32	Amorphous Silicon Module on Glass	35
Figure 33	a-Si Modules Built Into Roof Shingles	36

Figure 34	Representative Annual Change in Efficiency with Amorphous Silicon Modules	37
Figure 35	Large Mirror Concentrator. The Cells are Located at the Bright Spot	39
Figure 36	Mirror Augmented Concentrator	40
Figure 37	Grid-Connected Inverter (on right) with Disconnect (in center) and AC Load Center (on left)	44
Figure 38	Sine Wave Voltage from a 120-Volt Inverter	45
Figure 39	A Stand-Alone Inverter	46
Figure 40	AC Modified Square Wave Voltage from 120-Volt Inverter	47
Figure 41	Hybrid Inverter	48
Figure 42	Efficiency Curves At Different Array Voltages for a Certain Inverter	49
Figure 43	Battery Room, Coiba National Park, Panama, 24 Volts, 1,000 Ah	58
Figure 44	Battery Cap That Recombines Hydrogen and Oxygen to Re-Form Water	59
Figure 45	AGM Lead-Acid Battery	60
Figure 46	Gel Cell Battery	61
Figure 47	Battery Wiring Configurations	63
Figure 48	Basic Charge Controller	70
Figure 49	Maximum Power Point Tracking Charge Controller with Three-Stage Charging	71
Figure 50	Pulse Width Modulation	72
Figure 51	Typical Charge Controller Efficiency Curves for Different Input Voltages	76
Figure 52	Heads in a Water Pumping System	81
Figure 53	Solar Water Well System	82
Figure 54	Spring Box	84
Figure 55	Submersible Pump with Motor	86
Figure 56	Surface Pumps	87
Figure 57	Solar Submersible Horizontal Axis Piston Pump	88
Figure 58	Flexible Impeller Pump, No Motor	88
Figure 59	Jack Pump for Water Well in Heber, Arizona	89
Figure 60	Helical Rotor Pump, No Motor	90
Figure 61	Three-Stage Centrifugal Pump, No Motor	91
Figure 62	Discharge Curves for a Solar Helical Rotor Pump	93

LIST OF ILLUSTRATIONS XIII

Figure 63	Discharge Curves for a Solar Centrifugal Pump	93
Figure 64	UV Water Sanitation Unit	95
Figure 65	Solar Powered Reverse Osmosis Water Maker Installed in a Boat	96
Figure 66	Solar Electric System Losses	102
Figure 67	Solar Pathfinder	104
Figure 68	Rebate Example for Arizona, Sulphur Springs Valley Electric Cooperative	110
Figure 69	Loan Example, Tennessee Department of Economic & Community Development	111
Figure 70	Grant Example, Bonneville Environmental Foundation	113
Figure 71	Typical Residential Roof Structure	132
Figure 72	Modules Designed to be Used Without Attachment to the Roof	133
Figure 73	Roof Mounting System	134
Figure 74	Ground-mounted Frame in Boulder, Colorado	136
Figure 75	Village Power Array Designed for Snow	137
Figure 76	Ratcheting Compound Action Crimpers	140
Figure 77	Heavy Duty Crimping Tool	140
Figure 78	Proper Module Grounding Hardware	143
Figure 79	Rodent Damaged Wires	145
Figure 80	Metal Wire Tray Used To Protect Wire	146
Figure 81	An Orangutan-Proof Fence at Camp Leakey, Borneo	147
Figure 82	Digital Multimeter	150
Figure 83	Disconnecting Modules in Series	152
Figure 84	Clamp-on Ammeter	153
Figure 85	Current Shunt Resistor, 50 mV at 200 Amps	154
Figure 86	Plane of Array Precision Spectral Pynanometer	155
Figure 87	Solar Cell Based Pynanometer	156
Figure 88	Thermocouple and Thermocouple Reader	157
Figure 89	Infrared Thermometer	158
Figure 90	IV and Power Curves of a Solar Cell	159
Figure 91	Setup For IV Curve Measurement	160
Figure 92	Curve Tracer	161
Figure 93	Hydrometer	165
Figure 94	Battery Monitor	166
Figure 95	Corroded Battery Connection	167

Figure 96	No Rain for a While; The Module on the Right Has Been Washed Regularly	168
Figure 97	Hosing Off Modules from a Moving Pickup Truck	169
Figure 98	Shadow of Satellite Receiver	171
Figure 99	The IV Curve of a 3 Parallel by 3 Series Array With One Failed Module	176
Figure 100	Shadowing To Find Failed Module	177
Figure 101	Testing A Diode	178
Figure 102	Lock-Out and Tag-Out	185
Figure 103	Battery Terminal Covers	190
Figure 104	Fall Arrest System	191
Figure 105	Fall Arrest System	192
Figure 106	Roof Jacks with Plank Installed	194

LIST OF TABLES

Table 1 An Example of Voltages for State of Charge 164
Table 2 Approximate State of Charge Versus Specific Gravity 165
Table 3 Effect on the Body From Different Amounts of Current 183

FOREWORD

More and more people are installing solar electric modules on their houses. Why is this? Concern over pollution and global warming, rising cost of utility bills, rising cost of oil, available government financial incentives, and historically low cost of solar modules are all factors.

In 1956, Bell Labs produced the first solar modules (see Figure 1).[1]

Figure 1 Bell Labs Engineer Testing a Solar Panel in 1954.[2]

In 1956, a one-Watt solar cell cost $300 while today they are a fraction of that cost. The first real market was for satellites, where there was no alternative (see Figure 2). As the infant industry started to mature and prices started to fall, terrestrial applications started to become interesting. In the beginning, solar energy was only economical for specialized applications, such as mountain top radio transmitters or remote cabins. Solar energy is now beginning to compete with local utilities. Depending on local utility costs, amount of sunlight available, and government incentives, the cost of a solar system can be paid back in a few years by greatly reducing or eliminating utility bills. After the equipment is paid

---

[1] http://www.californiasolarcenter.org/history_pv.html
[2] http://www.bell-labs.com/org/physicalsciences/timeline/span10.html

for, the electricity is free.[3] As the cost of solar equipment continues to decrease, and utility bills increase, the payback period will continue to get shorter. The solar industry is poised for explosive growth in the future.

Figure 2 1973 Satellite Solar Cell Used on Skylab Space Station.[4,5]

---

[3] Except for a small amount of operations and maintenance cost.
[4] http://www.collectspace.com/collection/artifacts_slp_solarcell.html
[5] Courtesy NASA

INTRODUCTION

WHY SOLAR

Solar electric modules are reliable, economic, quiet, non-polluting, and easy to maintain. Best of all they use renewable energy. Solar modules can provide all the electricity needed for an average house, cabin, yacht, motor home, or other application. Additional modules can be easily added when more energy is needed.

Solar modules are reliable

Current production modules are warranted for 25 years and there is no reason why they couldn't last 50 years or more. Solar modules contain no moving parts. With some exceptions, the physics of solar cells do not change over time. Today's solar equipment will continue to perform long into the future.

Solar modules are economical

Under the right circumstances, solar electricity today costs less than buying electricity from the utility company. For some applications, solar modules are the only economic choice.

Solar modules are quiet

Solar modules do not generate any noise.

Solar modules are nonpolluting

During use, solar panels produce no toxic gas, exhaust, green-house gasses, oil spills, or fuel spills.

Solar modules are easy to maintain

Solar modules can provide years of service with little attention.

Solar modules use renewable energy

The source of energy is sunlight, which is free and abundant and is not depleted by using it.

Solar modules provide energy independence

You can provide for your own electricity without depending on a utility company or a fuel supplier.

Solar systems are expandable

It is easy to add more modules as need arises or time and money are available.

THE PURPOSE OF THIS BOOK

After reading this book, you should have a basic understanding of what solar modules are, how they are used, their strengths and weaknesses, and what other equipment is needed to make a solar system work. The book includes chapters on installation, maintenance, sourcing equipment, financial incentives, and safety. This book is written to be easy to understand. Mathematical formulas and technical jargon have been avoided as much as possible.

The companion book, *Solar Design*, goes into detail on how to design a solar system. It includes valuable software to aid you in the design process. *Solar Design* is intended for the dedicated do-it-yourself person or a professional installer. It also includes detailed technical information on how solar cells work and how modules are made.

TERMINOLOGY

Scientists and engineers call solar electric modules "**photovoltaic modules**," or "**PV modules**" for short. This book uses the more common name of "**solar modules**." The book does not discuss solar thermal collectors that are used to heat water. Solar modules here mean solar electric modules.

Solar modules are the basic building blocks of a solar system. They are made from solar cells, which are joined electrically and laminated into an environmentally sealed unit. A solar panel is a collection of one or more modules. A solar array is a collection of one or more panels.

Introduction

If you run across a new term or want to refresh your understanding, consult the glossary at the end of this book. More information can be obtained through a search on the Internet or at your local library. References are made to sources on the Internet at the time the book was written.

Price and Model Numbers

The solar industry is changing fast. Companies are being formed, being bought by other companies, and metamorphosing constantly. New products are coming out every day. Any information on pricing, brand, or model numbers will be obsolete by the time you read this. For that reason you will generally not find pricing and model numbers in the book. However, you will find current information on how to find prices, brand, and model numbers and how to interpret the information. The book includes a chapter on sourcing solar equipment.

wires are exiting. Some modules just have the wires and no junction box. On the back is also a label which lists the output power.

SOLAR MODULES USE SUNLIGHT

Solar modules use sunlight to make electricity. When you put them in the sunlight they are "on." More sunlight makes more electricity. You can turn them off by covering them with an opaque material like black plastic. If you point them in the direction of the sun they will make more electricity than if you point them away from the sun. They do not store energy and only produce when in sunlight.

Shadows have a unique effect on solar modules. If you shade only one cell in a module, the whole module will produce almost no electricity. Shadows can kill the output of a solar system (see Figure 4).

Figure 4 Shadows Greatly Reduce Module Output.

This is one of the most important things to remember. Avoid placing the modules where they will get shadows. Commonly overlooked shadow sources are other houses, trees, standpipes, other modules, rigging, and TV or satellite antennas. Nearly all solar arrays on boats have shadows to some degree (see Figure 5). Even small shadows can severely reduce output. Also, clean off any bird droppings, leaves, or other debris that fall on your modules.

Solar Module Basics

Figure 5 Shadow-Producing Equipment.
Radar Dome on a Boat Arch Mounted Solar System.

One solution for the common problem shown in Figure 5 is to make a double layer arch, with the radar dome mounted on the lower level and the solar panels mounted on the top layer.

Solar modules do not have to be hot to work. In fact, an increase in temperature reduces the output of a module. Some modules are not as susceptible to high temperatures as others (see Chapter 3). However, it is always important to provide air circulation under the module to keep it as cool as possible (see Figure 6). The more space under the module the better, but in most cases 3 to 6 inches is enough.

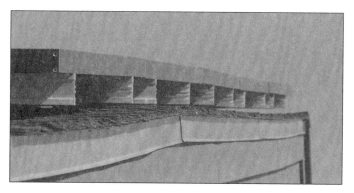

Figure 6 Ventilation Under Modules.

Solar modules produce less electricity when there is less sun due to clouds or haze in the air. You can even out the solar output over cloudy and sunny days by using batteries that store energy. Even so, you will need more modules if you are planning to use them in an area that has frequent clouds year round.

The output from your solar modules will also depend on where you use them (see Figure 7).

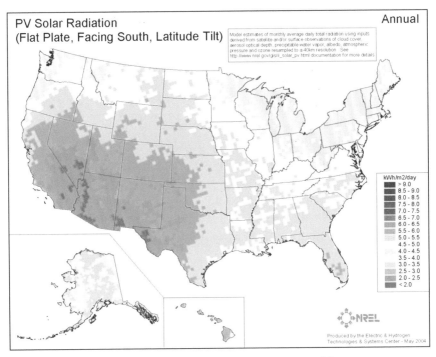

Figure 7 Annual US Solar Radiation.[7,8]

If you live in Arizona you will have twice as much annual output as you would if you live in Alaska. (Included with the companion book, *Solar Design* is a CD that has a database of many different sites in the U.S.)

---

[7] http://www.nrel.gov/gis/images/us_pv_annual_may2004.jpg

[8] To use the map, find where you live and match the color with the scale on the right. The numbers indicate the amount of sunlight falling per day at your location.

SOLAR MODULE BASICS 11

Module output also depends on the time of year. You can expect less output during the winter than the summer. For one thing, the days are shorter during the winter, and for another, there are more clouds. Figure 8 shows a map of solar radiation in January and Figure 9 shows the same map for July. You can get your own map for any month at **http://www.nrel.gov/gis/solar.html**.

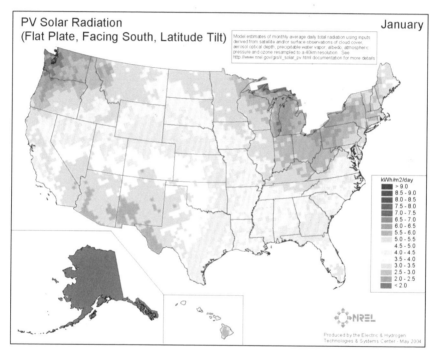

Figure 8 January US Solar Radiation.[9]

---

[9] http://www.nrel.gov/gis/images/us_pv_january_may2004.jpg

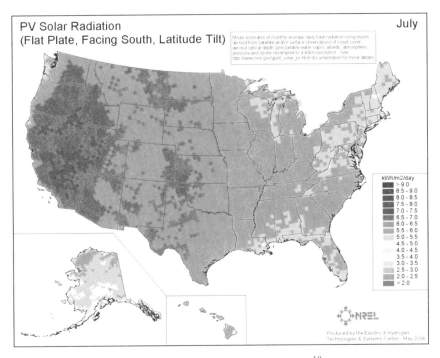

Figure 9 July US Solar Radiation.[10]

From the solar radiation maps, Hawaii looks like it has fairly constant radiation year round. This is generally true as you move closer to the equator. The days are more nearly equal in length all year. It also looks like it is impractical to try and get through the winter in Alaska with a solar system without a supplemental energy source. The days are short or nonexistent, and the sun is not very strong in Alaska in the winter.

Note that maps are for "Flat Plate, Fixed Tilt, Facing South."[11] This brings up the topic of orientation. If you face the solar modules towards the sun, you will get more output. In the northern hemisphere, this means facing south. It also means tilting the modules approximately at the same angle as your local latitude. In Raleigh, North Carolina, where the latitude is 36°,

---

[10] http://www.nrel.gov/gis/images/us_pv_july_may2004.jpg

[11] "Flat plate" is another name for the normal type of solar module.

fixed (stationary) modules should be tilted approximately 36° from horizontal for maximum yearly output (see Figure 10).

Figure 10 Flat Plate, Fixed Tilt, Facing South.[12]

(The companion book, *Solar Design*, includes software on the CD that allows you to optimize the tilt for your particular climate and application.)

You can point the solar modules towards the sun throughout the day, either by hand or by a machine called a tracker (see Figure 11). If you move the modules to follow the sun, you can get up to another 30% additional energy. Moving the modules by hand is inconvenient and won't work if you're not home. Boats sometimes use a system that allows the modules to be moved by hand. Trackers are expensive and contain moving parts that can contribute to higher maintenance costs. They are also hard to mount on roofs. The cost may not be justified except for large ground-mounted systems.

---

[12] Courtesy DOE/NREL, Carolyn A. Demorest

Figure 11 Suntracker Model 30 Tracker with Four Solar Modules Mounted.[13]

WHAT COMES OUT OF THE WIRES

Solar modules generate direct current (DC). The voltage can be anywhere from a few Volts to 50 Volts DC (VDC). Modules that are designed to charge 12-Volt batteries produce about 16 VDC. You can guess the voltage by adding up all the number of cells and dividing by 2, since each cell produces about half a Volt.[14] The best way to determine the voltage is to consult the label on the back (see Figure 12), or use a voltmeter. What you are measuring with the voltmeter is the Open Circuit Voltage (Voc).[15]

---

[13] Courtesy DOE/NREL, Glenn Elden

[14] Not always accurate

[15] Open circuit voltage is the voltage at the wires when the module is not connected to a load.

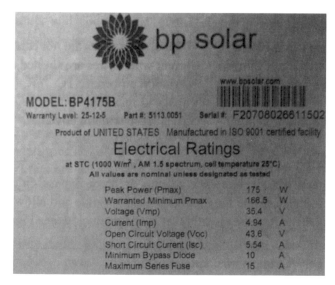

Figure 12 Typical Label on the Back of a Solar Module.

What happens when you short the wires together? Besides getting a spark, nothing. Unlike a battery, shorting the wires together will not harm a solar module. The current that flows when the wires are shorted is called the Short Circuit Current (Isc). The module is also not harmed if the wires are left unconnected.

What are those other numbers on the label? Peak Power (Pmax) is the maximum power that can be expected out of that module under Standard Test Conditions (STC). In normal operation you will almost never get this amount of power. That is because the solar radiation will normally be lower than the 1000 Watts per square meter (W/m^2) used for STC, the operating temperature will normally be higher than the 25°C used for STC, and the module will almost never be pointing directly at the sun. Expect 40% to 70% of the Peak Power.

The Peak Power occurs at somewhat less voltage than Voc and somewhat less current that Isc. The voltage and current at Peak Power are called Peak Power Voltage (Vmp) and Peak Power Current (Imp).

Battery systems for homes that are not grid-connected can be quite large (see Figure 22).

Figure 22 Off-Grid Solar Powered House.[28]

GRID-CONNECTED WITH BATTERY BACKUP

Grid-connected with battery backup is a combination of the grid-connected system and the battery system. A battery bank provides power for critical circuits when the utility company goes offline. The critical circuits power such things as the refrigerator, lights, telephone, security, and office equipment such as computers. When the utility company is offline, and while there is sufficient solar power, the loads are supplied and the batteries recharged by solar. There can be a generator to use if all else fails. These systems are normally used where there are frequent and prolonged power outages. When the utility company is online, the battery bank is charged by the utility power and any excess solar power.

Although it is possible to power the entire house during power outages using batteries and the solar array, this is usually too expensive for most homeowners. The important thing is to keep the critical systems going. The homeowner decides what is critical.

---

[28] Courtesy DOE/NREL, Solar Depot, Inc.

SOLAR POWERED WATER SYSTEMS

For solar powered water pumping, the solar array can be hooked directly to the water pump without a battery bank. The water source can be a well, stream, or pond. The water is usually stored in a water tank. If the water tank is located higher than where the water is used, normal water pressure is maintained even at night (see Figure 23). In some applications, a second pump and pressure tank are used to maintain water pressure. In this case a battery bank is used to run the second pump when the solar power is unavailable.

Figure 23 Solar Powered Water System with Storage Tank.[29]

A common application of a solar powered water system is to water livestock. The solar system replaces or augments the ubiquitous windmill (see Figure 24).

---

[29] Courtesy DOE/NREL, Kent Bullard - National Park Service

Figure 24 Solar Powered Livestock Watering System.[30]

Livestock watering systems are located in remote areas where utility power is unavailable. The water source is a usually a well which may limit the amount of water available. The system may also pump water from a stream to a trough so that the livestock do not contaminate the stream.

Solar powered water systems are also used to supply water for villages (see Figure 25). A well may be the only safe water source for a remote village and solar is usually the only practical power source to pump the water.

---

[30] Courtesy DOE/NREL, University of Wyoming

APPLICATIONS 29

Figure 25 Village Solar Water Pumping.[31]

For a remote residence, it is common to combine a solar powered battery system and a water pumping system. Since the battery is already there, a pressure tank can be used. If the battery system includes an inverter, an AC pump can also be used. The solar array is usually located close to the pump so that wiring losses are minimized.

OTHER APPLICATIONS

Solar energy is used in thousands of applications, from huge power plants to solar powered watches (see Figure 26). Most people are familiar with solar powered garden lights. More applications are being devised all the time. All solar systems are similar to what is discussed in this book. They all follow the same principles.

---

[31] Courtesy DOE/NREL, Cortina community water pump, MPPT, 1/2-hp DC Motor, Balanced-Beam Jensen Pump, 1000 Gallon Per Day, 150 feet Total Static Head, Cortina, California, Lloyd Herwig

CRYSTALLINE SILICON MODULES

Most of the modules sold today are crystalline silicon modules. Crystalline silicon has a regular molecular structure. These are the most efficient modules. The individual cells are bonded to glass and can be clearly seen. The color is dark blue to black with a white or black background. You may see a purple tint to the cells.

Crystalline silicon modules come in two basic forms: single crystal and polycrystalline. Single crystal silicon modules are made from slices of a single crystal of silicon (see Figure 28). The cells are usually square with rounded corners or they are round or half round.

Figure 28 Single Crystal Silicon Modules.[34]

---

[34] Courtesy DOE/NREL – Off-grid Photovoltaic Powered Home in Coal Creek Canyon Near Denver, Colorado, Dave Parsons

Types of Modules

Polycrystalline silicon modules are made from slices of polycrystalline material and are square (see Figure 29).

Figure 29 Polycrystalline Silicon Module.[35]

The individual crystals in the polycrystalline cell can be seen because they each reflect light differently (see Figure 30).

Figure 30 Polycrystalline Silicon Cells Showing Individual Crystals.

---

[35] Courtesy DOE/NREL – Solarex modules installed on the roof of the Satyanarayanpur Health Center in West Bengal, India, West Bengal Renewable Energy Development Agency

Because a-Si modules loose less power in high temperatures, they are sometimes preferred for hot climates, or for roof shingles that tend to get very hot. Because of their low-light performance, a-Si modules are found on indoor applications such as calculators and watches. Because of their flexibility, a-Si modules made on plastic are preferred for portable use such as military applications and backpacking.

Amorphous silicon cells are easily made with multiple junctions, which use more of the available sunlight and are therefore more efficient than single junction amorphous silicon cells. For more information on multiple junction cells, see the companion book *Solar Design*. Chapter 9.

Because of manufacturing reasons, thin-film modules may use non-tempered glass. Consult the manufacturer for the recommended loading for thin-film modules.

OTHER TYPES OF SOLAR CELLS

There are many other types of solar cells and more types being invented all the time. These specialty cells are not available commercially or are too expensive for general use. Gallium-arsenide (GaAs) cells are very efficient (and very expensive) and are used primarily for space applications. Crystalline silicon modules are the workhorse of the solar world and will likely remain so for a long time.

CONCENTRATOR MODULES

Concentrator modules use optical elements such as lenses or mirrors to concentrate the sunlight that falls on the cells (see Figure 35).

TYPES OF MODULES 39

Figure 35 Large Mirror Concentrator.
The Cells are Located at the Bright Spot.[42]

Solar cells put out current in proportion to the concentrated sunlight, up to the limit of the cell used. Concentrator modules must be used on trackers that

---

[42] Courtesy DOE/NREL, SAIC

follow the sun. The simplest concentrator is nothing more than mirrors placed on the sides of a regular module (see Figure 36). (Doing this may void the module warranty.) Concentrators such as these may not require a tracker.

Figure 36 Mirror Augmented Concentrator.[43]

Back in the days when solar cells were very expensive, replacing cell area with inexpensive optics and mounting the concentrators on trackers was a good idea. Now that the cost of cells has come down and the cost of the optics and tracking systems has increased, concentrators have taken more of a back-burner position. However, some people believe that given enough volume, concentrators could be cost effective for large-scale utility markets.

---

[43] http://www.iscat.com/newfiles/moltiplicatori.php Courtesy Centro Ricerche s.r.l.

TYPES OF MODULES

MODULE EFFICIENCY

Higher efficiency means more electrical energy will be produced for a given size array. High efficiency is important because less mounting structure, wire, roof area, land area, and other hardware are required. This can lead to a reduction in the overall cost of a solar system. High efficiency can be a deciding factor when trying to mount a solar system onto a limited space such as a roof, RV, or boat.

Module efficiency is easy to calculate. Divide the maximum peak power by the module area in meters and divide the result by 1000. For instance, a certain module has a peak power at standard conditions of 180 Watts and has an area of 1.33 square meters. The efficiency is 180/1.33/1000 = 0.135 or 13.5%. Remember that this is the efficiency at 25°C cell temperature and higher temperatures will mean less efficiency.

4
INVERTERS

In this chapter you will learn about inverters, what they do, where they are used, and the basics of how they work.

Inverters convert the direct current (DC) available from solar modules or batteries to the alternating current (AC) that most household electrical appliances use. The utility companies use AC rather than DC because it is more economical to generate and transport to the end user.

TWO BASIC TYPES OF INVERTERS

The two basic types of inverters are Grid-Connected and Stand-Alone. Grid-connected inverters, also called grid-tied, are used to supply power to the utility company and utility-fed AC loads. Stand-alone inverters, also called off-grid, are used to supply AC loads independent of the utility company.

GRID-CONNECTED INVERTERS

Grid-connected inverters have a special safety feature built into them. They are designed to sense the presence of the grid and shut down within a fraction of a second if the utility company goes offline. This means that an inverter designed for a grid-connected system will not work in a stand-alone system and will not supply power to your house in case the grid goes down. Grid-connected inverters are run directly from the solar array with no batteries in the system.

Grid-connected inverters are usually mounted close to the AC load center of the house. A disconnect switch is always placed between the inverter and the load center (see Figure 37).

Figure 37 Grid-Connected Inverter (on right) with Disconnect (in center) and AC Load Center (on left).[44]

The instant disconnect prevents "islanding." Islanding is when one or more grid-connected inverters continue to operate during a power outage. They can back-feed power to the grid, stepping up the voltage through the transformer. The linemen go out to fix the broken high-tension line only to find that there is still high voltage present. This can be dangerous, as you can imagine.

Grid-connected inverters produce exceptionally "clean" electricity, often cleaner than the utility itself. This means that the electricity is free from wave-form distortion, voltage spikes, frequency drift, voltage surges, and

---

[44] Courtesy www.pcitakoffs.com

other problems. Grid-connected inverters produce a sine wave voltage (see Figure 38). A pure sine wave voltage runs equipment more efficiently than other wave forms, causes less heating, and will not cause a hum in your stereo or interference on your TV.

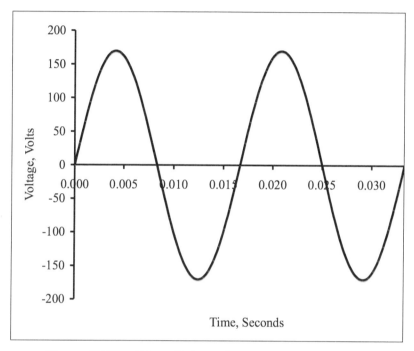

Figure 38 Sine Wave Voltage from a 120-Volt Inverter.

If you are planning a grid-connected system, the utility company will have to approve the inverter. Start working with your utility company as soon as you start to think about a grid-connected system. Inverters (and modules) approved by the California Energy Commission (CEC) are often accepted by utility companies in other states.[45]

Maximum Peak Power Tracking

A valuable feature built into grid-connected inverters is something called maximum peak power tracking (MPPT). MPPT runs the solar array at its

---

[45] http://www.gosolarcalifornia.org/equipment/index.html

most efficient operating point. Recall from Chapter 1 that solar modules have a current and voltage (Imp & Vmp) corresponding to the maximum power output. A MPPT inverter will constantly search for this point, which changes as sunlight and temperature vary. This assures that you get the best performance out of your modules, saving you money by reducing the required size of your solar array.

STAND-ALONE INVERTERS

Stand-alone inverters (see Figure 39) are designed to run off a battery bank that is charged with a solar array. They will run autonomously, not requiring the presence of the grid.

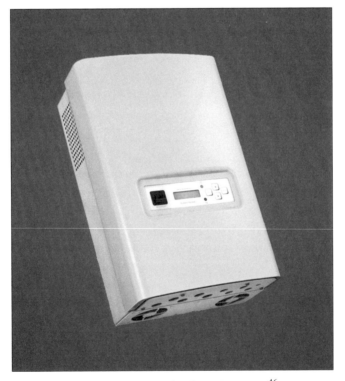

Figure 39 A Stand-Alone Inverter.[46]

---

[46] http://www.sma-america.com/ Courtesy SMA Solar Technology AG

NON SINE WAVE INVERTERS

Stand-alone inverters sometimes don't provide the perfect sign wave that grid-connected inverters do. Some stand-alone inverters have nothing more than a square wave or modified square wave (see Figure 40).

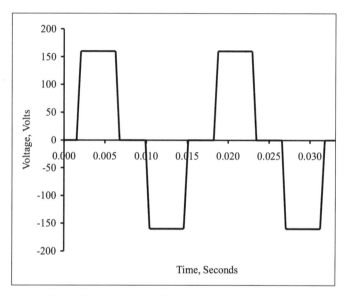

Figure 40 AC Modified Square Wave Voltage from 120-Volt Inverter.

Modified sine wave inverters approximate the sine wave, but have jagged components to their wave forms. These inverters are fine for some applications but may cause excess heating in motors or may ruin sensitive electronic equipment. They may cause a hum in audio equipment, interference in TVs, or cause surge protectors to not function. Consult the equipment and inverter manufacturers if you have concerns. Non sine wave inverters are lower cost than the pure sine wave inverters.

LOW VOLTAGE DISCONNECT

Since stand-alone inverters are run off of batteries, they will usually incorporate a low voltage disconnect (LVD) to protect the batteries. This feature will disconnect the battery bank when the voltage gets too low. The low voltage set point is adjustable. The inverter can be set to reconnect the battery bank automatically or manually when the voltage recovers.

HYBRID INVERTERS

Hybrid inverters are a combination of grid-connected and stand-alone (see Figure 41). They supply power to the utility, but when the utility goes down, they use a battery bank to supply AC to critical loads, such as a refrigerator, lights, or computer. Like grid-connected inverters, in a power outage these inverters automatically disconnect from the utility within a fraction of a second to prevent islanding. Hybrid inverters can have a battery charging circuit built in that will help to charge the batteries when the utility comes back online. While the power is out and there is sunlight, the batteries will be charged from the solar array. LVD may be used to protect the batteries. Hybrid inverters are used in areas with frequent or sustained outages or to protect critical equipment.

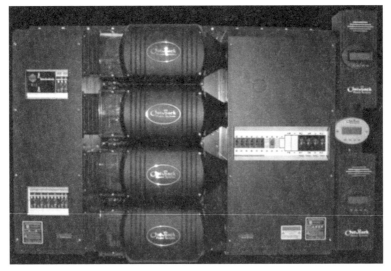

Figure 41 Hybrid Inverter.[47]

SURGE CURRENT

Some loads, such as certain electric motors, have a startup current far larger than the normal running current. This is called surge current or starting current. Inverters can handle a certain amount of surge current. The surge current can last for a few seconds and be several times the

---

[47] www.outbackpower.com

normal current. If you anticipate this type of load in your system, consult the inverter manufacturer and the company that makes the equipment you will be starting. It may be necessary to buy a larger inverter just for the surge current.

INVERTER EFFICIENCY

No inverter is 100% efficient. The efficiency of an inverter determines the amount of AC power that gets converted from DC power. Efficiency varies with the power the inverter is producing and the array voltage (see Figure 42).

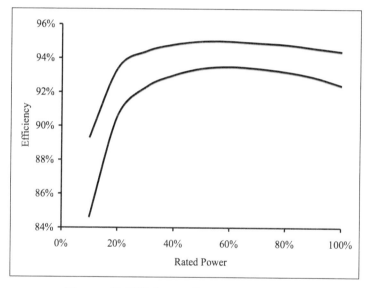

Figure 42 Efficiency Curves At Different Array Voltages for a Certain Inverter.

Inverter manufacturers quote maximum efficiency, but in normal use the efficiency is less. The California Energy Commission tests inverters and assigns a "weighted efficiency" to them, which is more realistic.[48] If available, use the weighted efficiency for design calculations. If no weighted efficiency is available, use an efficiency 2 percentage points

---

[48] http://www.gosolarcalifornia.org/equipment/inverter.php

lower than the manufacturers stated maximum efficiency for design calculations.[49]

High Temperature Derating

The inverter manufacturer will specify the operating temperature range of the inverter. Some inverters have a method of cutting back the AC power when their temperature gets too high. This protects the inverter, but means that it will not be delivering the power you expect on a hot day. This can be mitigated by mounting the inverter in a cool, shady spot, with lots of air circulation. Read the inverter specification and consult with the manufacturer if you have concerns.

Special Features

Inverters can come with additional features:

- Disconnects that satisfy code requirements, eliminating additional parts
- Outdoor enclosure allowing the inverter to be mounted outdoors
- Information displays that give production and operating information
- Remote monitoring and troubleshooting
- Data logging
- Built-in ground fault protection
- Surge protection
- Data exchange with other equipment

Stacking for 240 VAC

Some inverters can be stacked, meaning that two can be connected in series to provide a higher AC voltage. For instance, two 120-VAC inverters can be stacked to provide both 120 VAC and 240 VAC. Some inverters can also be connected in parallel to increase current capacity.

---

[49] 96% maximum efficiency becomes 94% for performance calculations

VA Capacity and Power Factor

AC systems can have inductive loads, such as certain motors, welders, or florescent fixtures. Inductive loads will cause current to flow in the wires that actually does no useful work.[50] This current is in addition to the useful current used to supply the power (Watts). The inductive loads store and release the current, like a spring. For instance, there may be 20 amps of useful current, but with the inductive loads added in, a total of 30 amps. Even though it does no useful work, the extra current loads up the inverter and can overheat it.

The power that does useful work is called the "real" power. The voltage times the current is called "apparent" power and is abbreviated as VA. The power factor is the ratio of real power to apparent power. A power factor of 1 means that all the current does real work, while zero means none of the current does real work.

Inverters with high VA capacity may be needed for AC systems with low power factors. You can also add devices to bring the power factor back towards 1, such as capacitors.

Ground Fault Circuit Interrupters

Ground fault circuit interrupters (GFCI) are devices that disconnect the circuits when current flows through the equipment grounding wires due to a ground fault. They may also be called a ground fault protection device (GFPD). This is similar to the ground fault interrupters (GFI) you likely have installed in some circuits in your house that shut off the current in case there is a short to ground. In the case of solar systems, the GFPD is designed to prevent a fire rather than protect you from being shocked. The National Electric Code (NEC) requires GFPD on any roof mounted solar array. Some inverters have these devices already installed, eliminating the need to buy and install a separate device.

---

[50] http://en.wikipedia.org/wiki/Power_factor

Battery Charger Built Into Inverter

It is common to have a battery charger built into an inverter meant for a stand-alone application such as an RV or boat. When AC power is available from a site hookup or shore power, the inverter charges the battery bank. When external AC power is removed, the inverter supplies AC power from the battery bank. Charging the battery bank from the solar array is a separate activity provided by a separate charger.

Display

Most inverters have a display of some sort. It may be a system of LED lights, or it may be an LCD screen. The display may just show that the inverter is working, or it may show how much energy was generated and used, and the net energy flow. Some show the avoided carbon dioxide generation. It may also show cost savings, voltages, currents, fault messages, frequency, harmonic component, and other information. The display may be built-in or remote. What display to use is a matter of personal preference.

Data Link

Inverters are available that transfer data to and from other devices. You can connect your computer to the inverter for up-to-the-minute reports or to setup the inverter. Inverters can be configured to communicate with other inverters or other equipment in the solar system.

NEC Required Features

The NEC requires certain over-current protection devices (fuses or circuit breakers), correct types of disconnects, and if the solar array is roof mounted, a GFPD. Some inverters come with the NEC required parts installed, and therefore save money, room, and installation hassles.

Certifications

Certifications from independent testing facilities are provided with inverters to document compliance with safety, quality, or performance requirements. Below is a short summary of the certifications usually provided with inverters.

UL 1741

Underwriters Laboratories Inc. A standard for inverters and charge controllers. Primarily assures that the equipment meets certain safety requirements.

CSA C22.2 No. 107.1

Canadian Standards Association. A standard for inverters and charge controllers. Primarily assures that the equipment meets certain safety requirements.

IEEE 1547

Institute of Electrical and Electronics Engineers. A standard for grid-connected inverters. Assures safety and performance of inverters designed to supply power to utility grids.

ETL Listed

Edison Testing Laboratory. Safety testing of electrical components. Assures that the inverter meets minimum safety standards.

ANSI/IEEE 62.41.2

American National Standards Institute. Specifies waveforms used to test surge performance of inverters and other electrical equipment.

FCC part 15 B

Federal Communications Commission. Deals with radio emissions of inverters and other electrical equipment. Limits the amount of interference to other equipment such as TVs and radios.

CAN/CSA STD. E335-1/2E

Canadian Standards Association. General safety standard for electrical appliances.

ISO 9001

International Organization for Standardization. A set of standards that assures quality control through record keeping, accountability, process monitoring, defect resolution, and other business practices.

HOW TO PICK AN INVERTER

Sizing an inverter is covered in the companion book, *Solar Design*.

The best inverter for a particular system depends on the application, the type and brand of modules, the wiring configuration of the solar array, and the features required. Call several solar equipment suppliers to get a list of inverters they recommend. Remember that they want to sell what they have in stock. Specifications and copies of the instruction manuals are available on supplier and manufacturer websites. Some module manufacturers have a list of inverters that they recommend or require for their brand. Certain modules require a positive ground, which is not available on all inverters. Review the specifications of the inverters you are considering. Look at cost, efficiency, features, voltage ranges, operating temperature range, mounting requirements, extra equipment you will need, and warranty. Make a checklist.

Solar magazines are good sources of current information about reliability and ease of use of inverters and other solar equipment. Solar expos offer face-to-face communication with vendors and manufacturers. See Chapter 10 for more information.

Kits

Solar systems are available packaged as a kit. The kit contains all the parts you need, including the inverters, solar modules, mounting hardware, disconnects, breakers, and equipment boxes. Consult with the equipment supplier if you are interested in a solar kit.

5
BATTERIES

In this chapter you will learn about batteries, the types of batteries, how to use them, and their limitations.

BATTERY BASICS

Batteries store electrical energy in the form of chemical energy. When a battery is providing electricity, there is a chemical change taking place in the battery.

Primary batteries, such as the ones you use in your flashlight, are not rechargeable, while secondary batteries are. There are a number of different types of secondary batteries, commonly called rechargeable batteries. Primary batteries are not in the scope of this book.

A cell is the technical name for a single electrochemical unit of a battery with one positive terminal and one negative terminal. The common AA battery is technically a cell, although in common usage it is called a battery. A battery is a collection of cells. The 12-Volt battery in your car has 6 cells. A battery bank is a collection of one or more batteries.

The capacity of batteries is rated in amp-hours (Ah).[51] A 100 Ah battery will supply 10 amps for 10 hours. Capacity changes with the rate of discharge. The same battery may supply 60 Amps for one hour (60 Ah) and 1.2 Amps for 100 hours (120 Ah).

The charging and discharging rate is designated by the C rate. The C rate is the current expressed as a fraction of the capacity of a battery in Amps. For instance, if you discharge a 100 Ah battery at 1C, you are discharging

---

[51] To calculate the Watt-hour capacity of a battery bank, multiply the Amp-hours by the Volts (Amp-hours X Volts = Watt-hours).

at 100 Amps. If you discharge it at C/10, you are discharging it at 10 Amps. The same goes for charging.

Depth of discharge (DOD) is the amount of energy you have taken out of the battery in percent of capacity. A 70% DOD means that 70% of the capacity has been removed and 30% remains.

TYPES OF RECHARGEABLE BATTERIES

Lead-Acid

Lead-acid batteries are the type used to start your car. They are the oldest, most developed type of commercial battery, and the lowest cost. Most solar applications use lead-acid batteries. Lead-acid batteries are heavy. Lead-acid batteries are damaged if completely discharged or overcharged. They are toxic and present an environmental hazard if not recycled properly. The battery charger, called a charge controller, can be very simple or a more complex design. Charger controllers are covered in Chapter 6.

Other Rechargeable Batteries

A number of other rechargeable battery technologies have been developed, some of which are sometimes used for solar applications. The major types are briefly mentioned here, but are not discussed in detail in the book.

Nickel Cadmium, Nickel Metal Hydride

Nickel cadmium (NiCad) and nickel metal hydride (NiMH) batteries store more energy for their weight than lead-acid batteries but are more expensive. They can supply high current during discharge. Special chargers are needed and they can be damaged if overcharged. NiCad batteries are not damaged if completely discharged but NiMH can be permanently damaged if completely discharged. NiCad batteries are toxic and present an environmental hazard if not recycled properly. NiMH batteries are less toxic than NiCad batteries.

BATTERIES

Lithium-ion, Lithium-ion Polymer, Lithium Sulfur, Lithium Iron Phosphate

This is a general class of battery based on lithium-ion exchange. The technology is fairly new. They have a high energy to weight ratio and low self-discharge rate. They can be damaged or explode if mistreated and take special chargers. This class of battery is used in consumer electronics, model cars and planes, battery powered tools, and hybrid passenger cars. These batteries are more expensive than lead-acid batteries. They have a limited lifetime, unrelated to the number of discharge/charge cycles.

Nickel-Iron

Nickel-iron (NiFe) batteries are extremely long life and low maintenance. They can tolerate lots of abuse, such as overcharging and complete discharging. NiFe batteries are occasionally used in solar powered battery systems, but they are less efficient and cost more than lead-acid batteries.

TYPES OF LEAD-ACID BATTERIES

All lead-acid batteries have the same chemistry, but different construction is available to suit different applications.

Flooded Cell Lead-Acid Batteries

Flooded cell batteries are the type most often used for starting vehicles. Flooded lead-acid batteries contain liquid sulfuric acid in a plastic case. They are higher maintenance than other types of lead-acid batteries but lower cost. Each cell is topped with a vented cap. When charging, flooded lead-acid batteries generate a small amount of hydrogen and oxygen gasses which escape through the vent.[52] The hydrogen and oxygen gas comes from the electrolysis of the water in the acid which must be replenished with distilled water on a regular basis. Flooded lead-acid batteries used in solar

---

[52] Hydrogen is explosive in air and oxygen. If the battery box or battery room is properly vented, the hydrogen escapes before there is any problem. Hydrogen is a very light gas and escapes upward. The vent should be at the top of the box or room. More on battery safety in Chapter 15.

systems sometimes have a clear plastic case that allows easy inspection. They also have a large volume of acid over the plates so that water does not have to be added as frequently. Flooded cell lead-acid batteries are the most prevalent type of battery used in solar systems (see Figure 43).

Figure 43 Batteries, Coiba National Park, Panama, 24 Volts, 1,000 Ah.

You can purchase special caps for vented batteries that recombine the hydrogen and oxygen to re-form water (see Figure 44).

Figure 44 Battery Cap That Recombines Hydrogen and Oxygen to Re-Form Water.[53]

The caps have a catalyst in them to reform the water. Remove the caps if you do an equalization charge (see Charging a Lead-Acid Battery below). Check with your solar equipment supplier or battery supplier to obtain these caps.

Valve Regulated Lead-Acid Batteries

Valve regulated lead-acid (VRLA) batteries have a pressure valve in the vent to prevent escape of the hydrogen and oxygen gasses, which then recombine to re-form water. If over charged, the pressure valves can still vent, and there is usually no way to replenish the water. VRLA batteries are considered maintenance free. Special charging voltages should be used on VRLA batteries to prevent venting.

---

[53] http://www.rollsbattery.com/

Absorbent Glass Mat

Absorbent glass mat (AGM) lead-acid batteries are a type of VRLA batteries where the acid is held in a fiberglass mat between the plates (see Figure 45).

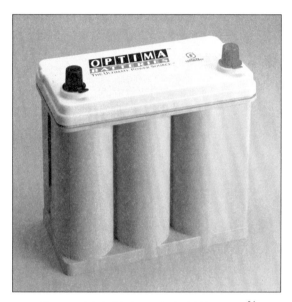

Figure 45 AGM Lead-Acid Battery.[54]

Compared to flooded lead-acid batteries, AGM batteries have higher energy per unit weight and resist shock and vibration better, but are more expensive. AGM batteries can be considered sealed and maintenance free under normal conditions. They can be charged and discharged at a higher rate but should not be overcharged. Special charging voltages should be used.

---

[54] http://www.absbattery.com/

Gel Battery

Gel batteries (gel cell) are VRLA lead-acid batteries in which the acid is gelled (see Figure 46).

Figure 46 Gel Cell Battery.[55]

Gel cell batteries are more shock and vibration resistant than flooded cell lead-acid batteries, but more expensive. The main advantages are low maintenance and very little chance of an acid spill. Gel cell batteries can be considered sealed and maintenance free under normal conditions. Special charging voltage and current are required.

---

[55] Courtesy Damon Hart-Davis, http://d.hd.org/

STARTING BATTERY VS. DEEP DISCHARGE BATTERY

Starting batteries are designed to start engines. They supply a large amount of current for a short time and are rated in cranking amps (CA) and cold cranking amps (CCA). They are only good for a 20% DOD, which means that a 50 Ah starting battery will only supply 10 Ah (50 Ah X 20% = 10 Ah). <u>Do not use starting batteries for a solar system.</u>

Use deep discharge batteries for solar systems. Deep discharge batteries are designed to be deeply discharged at lower current than starting batteries. They can be discharged to 80% DOD. A 50 Ah deep discharge battery will supply 40 Ah (50 Ah X 80% = 40 Ah), which is 4 times what the same size starting battery will supply. However, the deep discharge battery will not supply the huge amount of Amps that a starting battery will. If you have very large deep discharge batteries, for instance the house batteries on a large cruising boat, they can be used to start an engine. Consult the battery manufacturer and the engine manual.

Some batteries are advertised as being hybrid deep discharge starting batteries. These are a compromise between the two types. Unless you have a need for this kind of battery, do not use it in a solar system.

The battery manufacturer will supply you with the recommended charging and discharging rate, the recommended DOD, the expected lifetime, and the Ah capacity for their batteries.

BATTERIES

WIRING

Batteries are wired in parallel to add current and in series to add voltage (see Figure 47). This applies to solar modules and other equipment too.

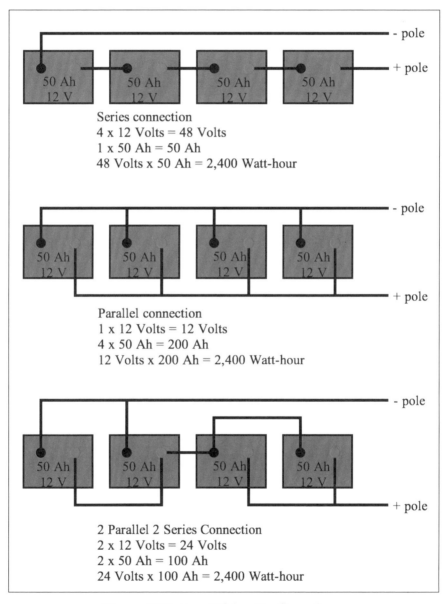

Figure 47 Battery Wiring Configurations.

EXPENSE AND MAINTENANCE

Over the life of a solar powered battery system, the batteries will be one of the most expensive components. This is not only due to the initial cost, but maintenance and replacement cost. The batteries will last 1 to 25 years, depending on their design, construction, and how they are treated. Of the many kinds of batteries, flooded lead-acid batteries are normally used because of their cost. VRLA batteries are sometimes used on battery backup systems because of their low maintenance.

Fully discharging a lead-acid battery will shorten its life considerably. The less deep the batteries are discharged, the longer they will last. Normally the batteries in solar powered battery systems are not discharged beyond 50% of their capacity.

In domestic solar powered battery systems the batteries are stored in a locked separate building or in a locked room away from the living space. See more about battery installation in Chapter 12.

The large batteries used in solar powered battery systems are usually made up of individual cells, which generate 2 Volts each. The 2-Volt cells are wired in series to produce 12, 24, or 48 Volts. If needed, the series strings are wired in parallel to add capacity. The battery case should have extra space below the plates to contain sloughed off material and extra space above the plates to hold the acid so that water does not have to be added as often.

Batteries can be very dangerous. Battery safety is discussed in detail in Chapter 15. Battery maintenance is covered in Chapter 14.

Batteries should be matched in a battery bank. Always use the same type, brand, capacity, and age of batteries to build a battery bank. This is because a weak battery can limit the capacity of series connected batteries. The weak battery can become severely discharged or reverse polarized, causing permanent damage. In a parallel connection, a weak or damaged battery can drain current from the other batteries, even to the point of

BATTERIES

causing overheating and a fire. For safety reasons, current limiting devices (fuses or circuit breakers) are required on parallel connected batteries.

Battery banks should be properly monitored and protected with low voltage disconnects (LVD). Instrumentation is discussed in Chapter 13. LVDs are discussed in Chapters 4 and 6.

Given proper care, batteries can last many years and be a reliable part of your solar system. A battery bank is one area where it pays to be conservative. Usage tends to grow and extended periods of low charging are always possible.

CHARGING A LEAD-ACID BATTERY

Constant Voltage Charge Controller

As a lead-acid battery is charged, the voltage increases. This allows the use of a very simple charge controller, one that supplies a constant voltage, about 13.7 Volts for a 12-Volt flooded lead-acid battery. When the battery becomes charged the increased battery voltage causes the current from the charger to diminish. This is called constant voltage charging and is used in most vehicles.

Three-Stage Charge Controller

The problem with constant voltage charging is that the battery does not get charged at the optimum rate. A discharged battery can accept a much higher current, which is not supplied by the constant voltage charger. To solve this problem, engineers have developed what is called a three-stage charge controller.

A three-stage charge controller charges differently depending on the state of charge of the battery. The first stage is called bulk charge. This is the highest current. The charge controller will determine when it is time to switch to the second stage, which is called the absorption charge. The absorption charge has a gradually decreasing current. The controller determines when to switch to the third stage, which is called float charge.

Float charge is the lowest voltage and is designed just to top off the battery and maintain a full charge. The float charge voltage is similar to the constant voltage charger. If needed, the charge controller will go back to the acceptance charge and then return to the float charge.

The voltages for the three charges, bulk, acceptance, and float, depend on the battery construction: flooded, VRLA, AGM or gel cell. The user can change the settings or order the charge controller set up for the type of battery used. Three-stage charging will provide a faster, more thorough charging of the battery bank. The battery manufacturer should be consulted to determine the maximum current and proper voltage for the particular battery.

Equalization Charge

Charge controllers can also supply a fourth type of charge, which is called equalization. This is a controlled overcharging of a flooded cell battery to remove contaminates from the surfaces of the plates. This provides a fresh surface, restores battery performance, and makes all the cells equal in voltage and state of charge. However, it does present two potential problems. First, the sloughed off material can settle in the bottom of the battery and short the plates. Second, equalization removes material from the plates, which can shorten the battery lifetime. Do not equalize a VRLA, maintenance free, gel cell, or AGM type battery as it can harm the battery. Consult the battery manufacturer if you have questions.

BATTERY TEMPERATURE COMPENSATION

Most charge controllers offer a way to measure battery temperature and adjust the charging voltage. For a 12-Volt battery, the charging voltage should decrease 0.15 Volts per every 10°F rise in battery temperature. In other words, if the float voltage is 13.7 Volts at 77°F, decrease the voltage to 13.55 Volts if the battery is at 87°F and increase it to 13.85 Volts if the battery is at 67°F. Charging at the correct voltage for the battery temperature can prolong the life of the battery and prevent thermal runaway.

BATTERY DESULFATOR

You can buy a device that is reported to convert lead sulfate crystals into the amorphous form of lead sulfate. Lead sulfate crystals form if the battery sits in a discharged state. They are not usable by the battery and are one of the things that an equalization charge removes from the surface of the plates. The amorphous form of lead sulfate is usable by the battery. Battery desulfators are designed to extend the life of a battery and even bring back a dead battery. If it works as advertised, it may eliminate the need for the equalization charge. Consult your solar equipment retailer if you are interested in trying these devices.

AGING

As lead-acid batteries go through discharge and charge cycles, they lose capacity. The loss is a gradual decline except at the end of life where there is a steep fall off. Voltage will also change. The life of a battery is determined by the design, but more importantly by how they are handled. More batteries are killed by the operator than from any other cause. To determine the expected battery life, consult the battery manufacturer. How to extend your battery life is discussed in Chapter 14.

ADDITIVES

Various battery additives on the market claim to extend the battery life or bring back a sick battery. Battery experts question the effectiveness of additives. If you are interested, consult the battery manufacturer to see what they recommend.

TEMPERATURE

Do not let your lead-acid battery freeze. If the battery is fully charged, the temperature it takes to freeze is very cold (-92°F), but as it becomes discharged if will freeze at a higher temperature (-16°F at 40% state of charge). A completely discharged lead-acid battery will freeze at 20°F. If frozen, do not try to charge, and once thawed, check the battery for cracks and leaks.

Lead-acid batteries start to rapidly lose capacity if colder than 50°F. They have more capacity if hot but will not last as long. The best temperature is room temperature, 50°F to 80°F. Monitor the battery temperature.

How To Choose Batteries

The companion book, *Solar Design* goes into detail on how to size battery banks.

The primary thing to consider is the capacity needed. You also have to choose the construction type, whether vented or sealed (VRLA). The lifetime of the battery is important and is related to the construction and depth of discharge. Budget constraints will have to be considered.

Wiring configuration is related to system size. Usually large systems are wired to provide 48 Volts. Small systems are usually 12 Volts and medium systems 24 Volts. Higher voltages are more efficient, and thus cost less, because the wire and equipment can be smaller.

Do not simply tap a high voltage battery bank to provide a lower voltage, for instance tapping 6 cells for 12 Volts from a 48 Volt or 24 Volt battery bank. This will unbalance the battery bank, reduce its capacity, reduce its life, and it may become a safety issue. Instead use a DC to DC converter. Another option for a 12 V / 24 V system is to use something called a battery equalizer. The equalizer will assure that the battery voltages will remain equal even though 12 Volts is drawn from half the bank.[56]

Sources for batteries are discussed in Chapter 10.

---

[56] http://www.vanner.com/vp/battery-equalizers.htm

6
CHARGE CONTROLLERS

In this chapter you will learn about charge controllers, what they are, what they do, and the various features available. This chapter only discusses charge controllers for lead-acid batteries. Charge controllers for other types of batteries are beyond the scope of this book.

THE FUNCTION OF A CHARGE CONTROLLER

The primary function of a charge controller is to control the charge to the battery bank. Charge controllers are only used on systems that have batteries. If you have a grid-connected system without battery backup, you do not need a charge controller. Likewise, if you have a solar array directly connected to a load, like a water pump, you do not need a charge controller.

As mentioned previously, a direct connection between a solar array and a battery bank is not recommended. This is a safety issue. The battery bank can be overcharged, which may cause overheating and, although remotely possible, a fire. Another more likely safety issue is that the battery bank can be a current source, supplying power to the solar array and creating a possible fire there or in the wiring. If you do install a solar panel that is directly connected to a battery, install an appropriate current limiting device (fuse or circuit breaker) and install a blocking diode to prevent current from flowing from the battery to the solar array.[57] Monitor the battery frequently and use the smallest size panel that will keep the battery charged.

---

[57] A blocking diode is a diode installed in the wire leading from the solar array to the battery. It is installed so that it allows current to flow from the array to the battery but prevents current flowing in the opposite direction. Size the diode at least two times the Isc and two times the Voc. Mount them according to the instructions, including a heat sink if called for. Blocking diodes are not normally needed if a charge controller is used.

TYPES OF CHARGE CONTROLLERS

Simple Charge Controllers

Charge controllers come with different features. The simple ones are the least expensive and just regulate the voltage to the battery (see Figure 48). Simple charge controllers charge batteries at a constant voltage designed to prevent damage from overcharging. Most simple charge controllers have built-in circuitry that eliminates the need for a blocking diode. A simple charge controller is inexpensive and greatly preferred over a direct connection between a solar module and a battery.

Figure 48 Basic Charge Controller.[58]

Three-Stage Charging

Advanced charge controllers have what is called a three-stage charging cycle which was discussed in Chapter 5. These charge controllers are more expensive but can save you money by doing a more efficient job of charging your battery bank, thereby reducing the required size of the battery bank and solar array.

---

[58] http://www.morningstarcorp.com/products/SunGuard/info/SG_DataSheet.pdf

MPPT Charge Controllers

A third type of charge controller combines the three-stage charging with maximum peak power tracking (MPPT) (see Figure 49).

Figure 49 Maximum Power Point Tracking Charge Controller with Three-Stage Charging.[59]

Inverters are also supplied with MPPT as was discussed in Chapter 4. A MPPT charge controller is the most expensive but is the most efficient use of the solar array. Although more expensive, MPPT can save you money by reducing the amount of solar modules you need.

---

[59] MPPT charger on a Boat, http://www.blueskyenergyinc.com/

Pulse Width Modulation

You will see the term PWM in relationship to charge controllers. It stands for pulse width modulation. This is a system where the voltage oscillates between zero and the maximum voltage available (see Figure 50).

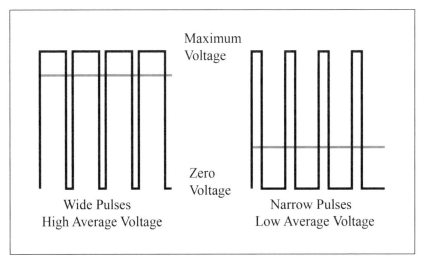

Figure 50 Pulse Width Modulation.

The width of pulses determines the average voltage that the battery sees and thereby the current. This is a similar technology to the dimmer switch that you use to control the lights in your house. It is more efficient than other methods to regulate the voltage. Some believe it does a better job of charging the batteries too.

Diversion Charge Controllers

One problem with a solar powered battery system is that there is a limited amount of energy that the battery bank can store. If the battery bank is already charged, any excess solar energy is wasted. A diversion charge controller solves this problem by diverting the excess energy to do useful work, such as heating water, heating a house, or running an irrigation system. The water heater could be a pre-heater for the normal water heater, or for heating water for a swimming pool or hot tub. Replacement heating elements for electric water heaters are available to match the voltage

CHARGE CONTROLLERS 73

available from the charge controller. If you are using a wind or water turbine to augment the solar array, a diversion charge controller may be necessary to keep the turbine from over-revving. Consult the turbine manufacturer. A diversion charge controller may be needed in addition to the charge controller normally used to charge the battery bank.

MULTIPLE SOURCE CHARGE CONTROLLERS

Multiple sources of energy, such as solar, wind, hydro, and a generator, will increase the reliability and flexibility of a battery powered system. Some charge controllers are designed to accept power from different sources, including a fuel cell.

A solar system and a standby generator can be synergistic. The solar array can be smaller and less expensive. The generator will be more economical and last a longer time if only used occasionally to charge the batteries or run a large load. Some charge controllers can be configured to automatically start the generator when the battery voltage is too low. Consult with the charge controller manufacturer if this is the kind of system you are considering. Wind, hydro, and engine driven generators are beyond the scope of this book.

CHARGE CONTROLLER RATINGS

Charge controllers are rated by the following:

- The maximum current they supply to the battery bank, Amps
- The nominal battery bank voltage, usually 12, 24, 36, or 48 Volts
- The maximum solar input power, Watts
- The solar input voltage range, Volts
- The maximum solar input current, Amps

The array voltage normally has to be higher than the battery bank voltage. Charge controllers specify the required array voltage corresponding to a nominal battery bank voltage.

TEMPERATURE DERATING

Like inverters, advanced charge controllers will start to cut back on the delivered current as they get hotter. This protects the charge controller from damage. The manufacturer will specify the operating temperature range of the charge controller and at what temperature the controller will start to cut back current. If you live in a warm climate or will have the charge controller exposed to high temperatures, this is something that may determine which controller you pick. If possible, mount the controller in the shade or in a cool part of the house.

TEMPERATURE COMPENSATION

As discussed in Chapter 5, temperature compensation is a method to change the charging voltage according to the temperature of the batteries. Unless your battery bank is always at the same temperature, this is an important feature. Temperature compensation is available on most charge controllers.

CHARGING VOLTAGE

The exact charging voltage depends on the type of battery: flooded, VRLA, AGM, or gel-cell. Unless the charge controller is a very simple type, it will have a way to adjust the charging voltage for the type of batteries you are using. The voltage will be different for each stage in a three-stage charge controller and will be different again for an equalization charge. Consult the battery manufacturer for their recommended charging voltages. Do not equalize sealed batteries.

LOW VOLTAGE DISCONNECT

Like inverters, charge controllers can have a low voltage disconnect (LVD) that disconnects the battery bank when the voltage is low. The LVD can be configured in various ways by the user.

Some charge controllers can be used as either an LVD or a charge controller. This means that you will have to buy two units. Other charge controllers have the LVD built into them so that the load current is run back through the charge controller, in which case you will only have to buy one unit.

HIGH VOLTAGE DISCONNECT

A high voltage disconnect (HVD) disconnects the solar array when its voltage is too high. This protects the charge controller from being damaged. High solar array voltage can occur during cold weather and low levels of sunlight.[60] The HVD will reconnect once the solar array voltage returns to normal. The HVD can be a separate piece of equipment or built into the charge controller.

PARALLEL CONFIGURATION

Most charge controllers can be wired in parallel to provide more charging current. Breaking up the charging task to two or more charge controllers may be an advantage if you have two or more solar panels that are mounted on different roof slopes or different roof orientations, or if you have different types of modules in each panel. For instance, take the case of two panels, one mounted on an east-facing slope and one panel mounted on a west-facing slope. The east-facing panel will have more current in the morning and the west-facing panel will have more current in the afternoon. By using two three-stage, MPPT charge controllers for the two panels, you can increase the total energy production. The controllers have to be linked to assure that they are both at the same charging stage.

---

[60] Lower temperature raises array voltage. Low levels of sunlight do not heat the modules as much as higher levels of sunlight. Although it may seem counterintuitive, cold weather enhances module output.

EFFICIENCY

Charge controllers have efficiency curves similar to inverters (see Figure 51).

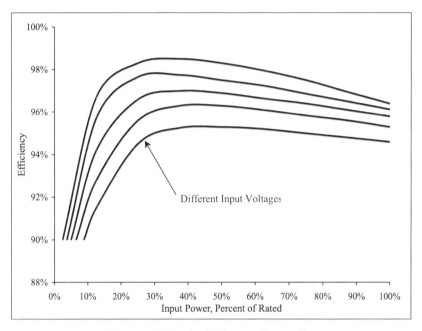

Figure 51 Typical Charge Controller Efficiency Curves for Different Input Voltages.

The efficiency depends on several factors, including battery voltage, input voltage, and the input power. Charge controller manufacturers will quote the efficiency at some point, usually the maximum efficiency. Due to variation in array output, the average efficiency will be less than the maximum. Consult with the manufacturer to get the efficiency curves. For calculation purposes, use an average efficiency two percentage points below the efficiency corresponding to your rated array voltage and power. For instance, if the efficiency is 96%, use 94%.

Efficiency of a charge controller is only a valid metric when the controller is operating at full capacity. After the battery bank is charged, it is the function of the charge controller to limit current, which greatly lowers the apparent efficiency.

Many charge controllers have a nighttime standby or sleep mode. This saves electricity by turning off the unit when it is not needed. A small amount of current drain is still there so that the controller will wake up when the solar array starts producing. The charge controller manufacturer will list the standby current if the controller includes a sleep mode.

GROUND FAULT PROTECTION DEVICE

Like inverters, charge controllers will need a ground fault protection device (GFPD) when the solar modules are mounted on a roof to prevent fires. The GFPD can be a separate device or built into the charge controller.

MOUNTING

Charge controllers are available to be mounted indoors or outdoors. If you want to mount an indoor charge controller outdoors, you will have to provide a weathertight enclosure. In any case, mount the charge controller in a cool dry location, where it will not receive direct sun. Charge controllers generate heat and must have airflow available. The charge controller will operate more efficiently and last longer if kept cool. Follow the manufacturer's instructions.

DISPLAY

Like inverters, charge controllers can have a variety of displays, from the most basic LEDs to LCD screens. The display can be built into the charge controller or mounted remotely.

DATA LINK

Also like inverters, data links are available for charge controllers. The data links can be used for programming, reporting, or interfacing the charge controller with other equipment, including other charge controllers.

CERTIFICATIONS

The certifications for charge controllers are much the same as those for inverters (see Chapter 4).

HOW TO PICK A CHARGE CONTROLLER

How to size a charge controller is covered in the companion book, *Solar Design*.

The charge controller is normally chosen after the battery bank and charging scheme have been chosen. The charge controller has to match the charging voltage and current of the battery bank. The required voltage and current for the charge controller will determine the wiring of the solar array.

The cost of the controller is important but not the only issue. Consider the features available, such as low voltage disconnect, high voltage disconnect, GFPD, type of display, data link, and temperature derating. Also consider the warranty, service, and support. Usually the controller that charges the most effectively is best because it will reduce the cost of the solar array.

Source for charge controller and discussed in Chapter 10.

7
SOLAR WATER SYSTEMS

In this chapter you will learn about solar powered water pumping, the uses, types, systems, and features. Sizing solar powered water systems is covered in the companion book, *Solar Design*.

TYPES OF SOLAR WATER SYSTEMS

Solar water systems are normally used for low volume water supply for domestic use or watering livestock. However, solar can also power a large irrigation system using a large solar array and a large pump.

The two basic types of solar water pumping systems are those with batteries or those without batteries.

Direct-Connect Solar Water Pumping

Direct-connect solar water pumping systems do not use batteries. The solar array is directly connected to a DC water pump, usually through a pump controller. They do not require an inverter, charge controller, batteries, or associated hardware.

An example is a water supply for a residence or village. In this example, a water storage tank holds the water so that it can be used at night or during cloudy weather. The tank is elevated to provide the necessary household water pressure, about 40 pounds per square inch (psi). The tank is sized to provide water for expected cloudy days and a reserve for fire fighting.

Another example is livestock watering. This is similar to the residential system except the tank, or trough, is mounted on the ground where the livestock can get to it. In areas were the sun may not shine for days, a separate storage tank is used so that the trough can be kept full.

Battery Powered Solar Water Pumping

Battery powered solar water pumping systems have a battery to extend the hours that pumping can be done. These systems are usually part of a larger solar powered battery system. The system may have two pumps, one to fill a storage tank and another to pressurize water in a pressure tank to supply the house.[61] The pump to fill the storage tank might be run only while the sun is shining.

TERMINOLOGY

In water pumping systems, pressure is referred to as "head" and is measured in feet or meters of water. With fresh water, it takes 2.31 feet of head to equal one psi of pressure. Static head is the pressure with the water not moving. If a pipe discharge is 40 feet above the outlet of a pump, the static head at the pump is 40 feet.

Dynamic head is the pressure due to the water moving in the pipes. It is caused by the friction of the water flow. Faster flow causes more dynamic head. Smooth pipes cause less dynamic head than rough pipes. Pipes tend to develop more dynamic head as they get older and full of corrosion and deposits on the inside. Pipe fittings have more dynamic head than the same length of straight pipe. A sharp elbow will develop more dynamic head than a gradual elbow. A larger diameter pipe will develop less dynamic head than a small pipe.

Total dynamic head (TDH) is the sum of static head, dynamic head, and suction head (see below). To size pumps you need to know the total dynamic head. Many online calculators are on the Internet for figuring total dynamic head.[62] Search for "TDH Calculator" or equivalent.

---

[61] A pressure tank is used to minimize the cycling of the pressure pump. Water is pumped into the pressure tank against air pressure in the tank. When the tank is full of water, the pump stops, and the compressed air supplies the pressure necessary to keep the water flowing. When the pressure tank is nearly empty, the pressure pump turns back on to start the cycle anew.

[62] http://www.southshoregunitepools.com/resources/calcs/TDHcalculator.htm, http://www.moneysaverpumps.com/calculator.htm, http://www. pumpworld.com/headcalc.htm

Suction head is on the inlet side of the pump. For instance, if the water level is 10 feet below the pump, the static suction head is 10 feet. In reality, the atmospheric pressure is pushing the water from the water surface to the pump. At sea level the atmospheric pressure is about 34 feet depending on the barometric pressure. The suction head that a particular pump can generate depends on its design, but is usually not greater than about 20 feet. It is less at higher elevation where the atmospheric pressure is less. If you have a pump located above the surface of the water, add the suction head when you calculate the total dynamic head (see Figure 52).

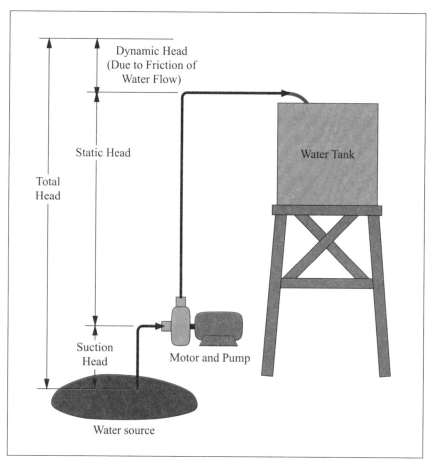

Figure 52 Heads in a Water Pumping System.

Except for mechanical strength considerations and the design of a particular pump, there is no limit to the amount of head on the pressure side of the pump.

SOURCES OF WATER

Wells

In developed countries, except for municipal water supplies, the most common source of water is a well. A well is nothing more than a hole in the ground to reach an underground aquifer, which can be an underground stream or an underground pool of water. The source of water can be many hundreds of miles away or many hundreds of years in the past. If it is not being replenished at the rate that it is being used, the water level will fall.

A modern well will have a plastic casing lining the well (see Figure 53).

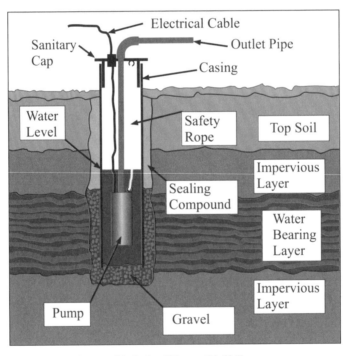

Figure 53 Solar Water Well System.

The casing protects the well from caving in and to seal it against groundwater. Groundwater is water that may have just fallen as rain or comes from a stream or other aboveground source. Groundwater is usually contaminated while the underground aquifer is usually not contaminated (not always). A seal on the outside of the casing prevents the groundwater from seeping down to the underground aquifer. The top of the casing is sealed with a cap to prevent contamination.

The well driller usually fills out a well driller's report and files it with the county. He will give a copy to the landowner. The report will detail the well construction, what depth it was drilled to, the flow rate and the water level at different flow rates. It may also include the type of strata drilled through and the material at the bottom of the well (gravel, sand, shale, etc.). Since wells do not supply an infinite amount of water, it is important to size the pump to match the production rate. It may also be necessary to match the type of pump to the quality of water. For instance, if there is a lot of sand in the water, make sure that the pump will not be damaged by sand and that appropriate screens are used. Consult with the pump manufacturer.

Water Springs

A spring is a place where the water in an underground aquifer is forced to the surface by the pressure in the aquifer. The pressure is caused by the water coming from a higher elevation. The aquifer may have reached a fault line that changed the direction of the water and forced it to the surface. Springs are usually tapped using a spring box which is a perforated plastic, wood, or concrete box buried in the spring and surrounded by gravel (see Figure 54).

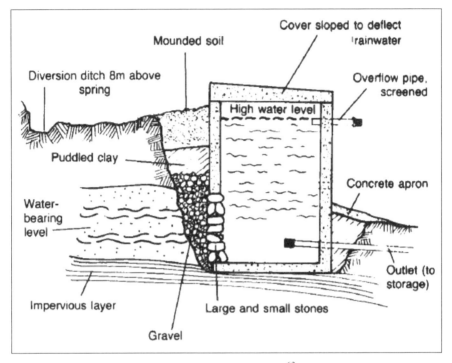

Figure 54 Spring Box.[63]

The water in the spring seeps into the spring box through the gravel and perforations. Because the spring box is buried, the water source is protected from surface contamination. The spring box usually feeds a tank through a pipe. The tank may be below where the water is to be used, such as a house. In this case, a solar powered pump can be used to transfer the water to a holding tank located higher than the house so that normal water pressure is available. Alternately, a solar battery system can be used to power a pressure pump with pressure tank.

Surface Water

A surface water source can be a lake, pond, river, or stream. A solar pump is used to transfer the water to a tank. In some cases, such as crop irrigation, the water is used directly without storage.

---

[63] http://www.lifewater.org/resources/tech_library.html Courtesy USAID Water For The World Technical Notes

Surface water is likely to be contaminated. If used for human consumption, it can be treated by various means including filtration, aeration, ozone, chlorine, heat, or UV light. Water treatment is discussed below. The surface water is sometimes tapped by drilling a shallow well next to the source. The ground provides natural filtering, but does not eliminate the need for testing and possible treatment.

Water Quality Testing

All water should be tested before being used for human consumption. It may contain dissolved minerals, heavy metals, arsenic, bacteria, or other contaminates. A testing service is usually available at a business that sells treatment products. Testing kits are available online. More information is available from the Centers for Disease Control and Prevention.[64]

TYPES OF PUMPS

One measure of a water pump is the amount of head it is designed to pump. A high head pump is used in a deep well and a low head pump is used for surface water or a shallow well.

The power needed to pump water is proportional to the flow times the head.[65] It takes the same amount of power to raise a small amount of water from a deep well as it does a large amount of water from a shallow well.

---

[64] http://www.cdc.gov/ncidod/dpd/healthywater/factsheets/wellwater.htm

[65] Neglecting friction, by definition one horsepower is equivalent to raising 33,000 pounds of water one foot in one minute, or one pound of water 33,000 feet in one minute. (horsepower = pounds of water per minute times rise in elevation in feet divided by 33,000) See http://en.wikipedia.org/wiki/Horsepower.

Submersible Pumps

Submersible pumps are designed to be lowered into the bottom of a well and used below the water surface (see Figure 55).

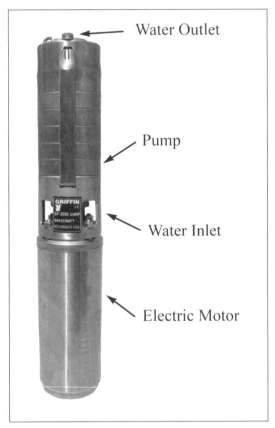

Figure 55 Submersible Pump with Motor.[66]

The motors in submersible pumps are sealed and located below the pumps. The electrical wires are sealed. A pipe, usually flexible plastic, connects to the top of the pump and supports it. A safety line is also attached to the pump in case the pipe breaks. A foot valve (check valve) at the bottom of the pipe prevents water from flowing back down the pipe when the pump

---

[66] http://www.griffindewatering.com Courtesy Griffin Pump and Equipment, Houston, Texas

is off so that the pump does not have to refill the pipe each time it turns on.[67] Submersible pumps can also be used to pump surface water but need to always be submersed in order to work. Submersible pumps are usually high head pumps but can be low head if designed for a shallow well.

Surface Pumps

Surface pumps are not submersible and must be protected from getting wet (see Figure 56). They are usually located above the surface of the water they are pumping and usually have to be primed. Priming is the process of filling the suction pipe with water. A foot valve is used to keep the suction pipe full. Surface pumps are usually low head pumps but can be high head if used for systems where the tank is located much higher than the source.

Figure 56 Surface Pumps.[68]

A special kind of surface pump is the jet-action pump. Part of the water is pumped down into the well to force more water to the surface by the action of a jet. These pumps are used for medium depth wells where the suction head is too great for a pure surface pump. The advantage is that the pump and motor are accessible at the top of the well. The disadvantage is that they are inefficient. Jet-action pumps are not recommended for solar systems.

---

[67] If the delivery pipe is subject to freezing, it is necessary to have a small "weep hole" below the frost line to drain water back into the well so that it does not freeze in the pipe. A column of water is still kept in the pipe below the weep hole.
[68] http://www.conergy.us/

Piston Pumps

Piston pumps use pistons in cylinders to pump the water (see Figure 57). Piston pumps are positive displacement pumps, meaning that they trap a fixed volume of water and then force that volume up the discharge pipe. Piston pumps are efficient and a good choice for a solar system. Piston pumps usually have multiple pistons.

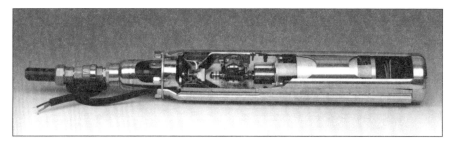

Figure 57 Solar Submersible Horizontal Axis Piston Pump.[69]

Flexible Impeller Pumps

Flexible impeller pumps are another type of positive displacement pump (see Figure 58).

Figure 58 Flexible Impeller Pump, No Motor.[70]

Flexible impeller pumps have a rubber rotor with a number of radial vanes. The housing is shaped to compress the vanes at the outlet, forcing the

---

[69] http://www.flowman.nl/ Courtesy of Fluxinos Italia, Grosseto, Italy

[70] http://www.duex-daesung.com/ Courtesy Pumps 2000 Europe LTD. ApS.

water out the pipe. Their main disadvantage is that the impellers wear out and need to be replaced periodically.

Jack Pumps

Jack pumps have the motor and gearbox located at the top of a well and a piston pump located at the bottom (see Figure 59).

Figure 59 Jack Pump for Water Well in Heber, Arizona.[71]

A rod connects the two and moves up and down with each stroke of the piston. An oil well pump is a jack pump in which the grasshopper-looking mechanism is the aboveground motor and gearbox. The old style windmill water pumps are jack pumps. The advantage is that the motor and gearbox are accessible while the actual pump is at the bottom of the well. The disadvantages are that it is hard to pull the pump for maintenance and these pumps are expensive. Modern, reliable submersible pumps are replacing jack pumps.

---

[71] http://www.sunpumps.com/

Helical Rotor Pumps

Helical rotor pumps are also called progressing cavity pumps (see Figure 60).

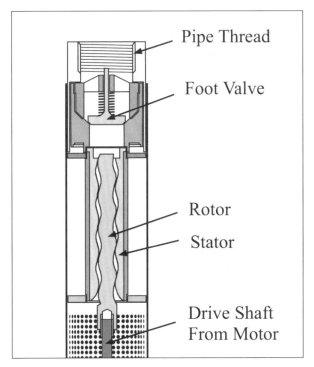

Figure 60 Helical Rotor Pump, No Motor.

Helical rotor pumps are positive displacement pumps. A polished metal helical rotor rotates in a rubber-lined housing (also called a stator), causing cavities between the rotor and housing to move axially and pump the water. There are no valves. Helical rotor pumps are high head pumps and used in submersible applications such as deep wells. They are efficient and reliable and good choices for solar power.

Centrifugal Pumps

Centrifugal pumps work by accelerating the water in a rotating impeller and then converting the kinetic energy to pressure (see Figure 61).

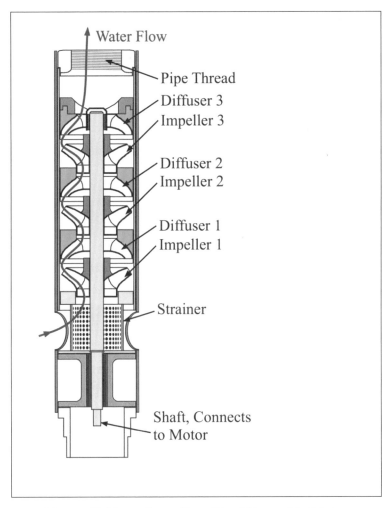

Figure 61 Three-Stage Centrifugal Pump, No Motor.

Centrifugal pumps are not positive displacement pumps and even at a constant rpm, the flow varies depending on the head. Centrifugal pumps can have one stage or multiple stages. A stage is one rotating impeller. Single stage pumps are low head and used for high volume pumping. Multiple-stage pumps can develop high head and are used for submersible pumps in medium to deep wells. Centrifugal pumps can be used for solar installations, with a slightly lower efficiency than positive displacement pumps.

Other Types of Pumps

There are many other pump technologies. One of the earliest pumps is the Archimedes screw described by Archimedes in 300 BC but first used in 700 BC. The surprising thing is that Archimedes screw pumps are still used today.

Some pumps are not appropriate for solar systems because they are not efficient. The solar modules are expensive and it pays to have the most efficient pump to minimize the overall cost and get the most water for a given size solar array.

Brush or Brushless

DC motors for pumps can have brushes or they can be brushless. The brushes are used to supply electricity to the armature in the motor, which is the rotating part. The problem with brushes is that they wear out and need to be replaced. Brushless DC motors eliminate the brushes by using electronics.[72] Brushless motors are more expensive but require less maintenance.

DISCHARGE CURVES

The pump manufacturer will supply discharge curves for a particular model of pump (see Figures 62 and 63).

---

[72] For more information about brushless DC motors, see http://en.wikipedia.org/wiki/Brushless_DC_motor

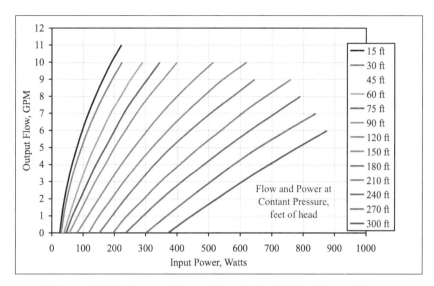

Figure 62 Discharge Curves for a Solar Helical Rotor Pump.

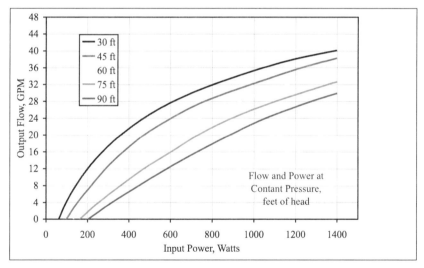

Figure 63 Discharge Curves for a Solar Centrifugal Pump.

The discharge curves can be used to find the expected discharge for a given head and power input or the required power for a given head and flow. For instance, the helical rotor pump in Figure 62 will discharge 9 gallons per minute (GPM) with a 90-foot total head and 350 Watts input.

The centrifugal pump in Figure 63 will discharge just under 6 GPM with the same 90-foot total head and 350 Watts input. (A 90-foot head and 350 Watts input is not the most efficient operating point for either pump.)

Efficiency

Efficiency is the fraction of power out divided by power in.[73] Using Figure 62 as an example, if the output is 9 GPM at 90-foot head with 350 Watts input, the power out is 0.204 horsepower (9 gallons per minute x 8.33 pounds per gallon x 90 feet / 33,000 foot-pounds per minute per horsepower). This converts to 152.5 Watts output (0.204 horsepower x 745.7 Watts per horsepower). So the efficiency at this operating point is 43.6% (152.5 Watts out / 350 Watts in). This is not the most efficient operating point for this pump. A higher efficiency of 47.1% is at a head of 150 feet, a flow of 7.5 GPM, and an input of 450 Watts. The pump manufacturer can advise you on the most efficient operating point.

CONTROL BOX

Except for very small systems, a control box is used to control power going to the pump. Control boxes will soften the startup so that there is less surge current, optimize the voltage to the pump, keep the solar array at its maximum power point (MPPT), shut off the pump when the tank is full, and perform safety functions like preventing the pump motor from being burned up from overload.

WATER TREATMENT

Solar energy can also be used to sanitize a domestic water supply. Several options are available.

The simplest option is to store water in PET plastic bottles (the kind of bottles you buy water in) in direct sunlight for a couple days. This is done in many developing countries and is the lowest cost method.[74] The

---

[73] The companion book, *Solar Design* includes a spread sheet on the CD that can be used to calculate pump efficiency

[74] http://en.wikipedia.org/wiki/Solar_water_disinfection

combination of ultraviolet (UV) radiation and heat from the sun, plus the dissolved oxygen in the water kills the bacteria in the water.

You can use the electricity from your solar array to generate your own UV (see Figure 64).

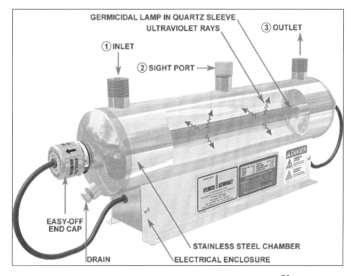

Figure 64 UV Water Sanitation Unit.[75]

The combination of heat and UV is highly effective.[76] Sealed units are available commercially. Search under "home UV water sanitation" or equivalent.

Another way to sanitize water is to bubble ozone into your water, powering the ozone generator with solar. This works like chlorine to oxidize any bacteria in the water and has the added advantage of reducing dissolved iron and minerals. Unlike chlorine, it does not leave an aftertaste. The ozone is formed from the oxygen in the air so that no chemicals need to be purchased. You can bubble the ozone into your storage tank and then take the water at a point above the bottom of the tank to avoid the sediment that forms, or you can bubble it into the water at the

---

[75] http://www.ultraviolet.com/

[76] http://www.who.int/water_sanitation_health/dwq/wsh0207/en/index4.html

outlet of a well and then remove the gas and filter the sediment. If the storage tank is located away from the house, a separate solar system is a good option. This contributes no harmful effects to atmospheric ozone.

If you need to treat your water source, consult with companies that provide this equipment.

Reverse Osmosis

Solar powered reverse osmosis is used to remove salt from seawater, especially on boats (see Figure 65).

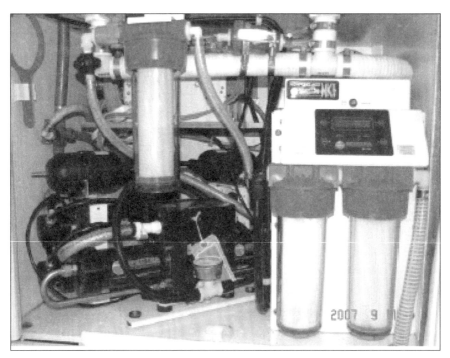

Figure 65 Solar Powered Reverse Osmosis Water Maker Installed in a Boat.

Reverse osmosis works by forcing water under high pressure through a membrane. The membrane acts like a filter allowing the water to pass but not the salt. The water that is produced from seawater tastes as good as expensive bottled water.

Reverse osmosis is not cheap, but sometimes it is the only domestic water available. Many areas of the world depend on reverse osmosis for their water and a lot of these systems are run by solar power. Lack of natural fresh water and lots of sunshine seem to go hand in hand.

HOW TO PICK A PUMP

Sizing a solar powered water system is covered in the companion book, *Solar Design*.

The important factors are durability, reliability, and efficiency. The cost of the pump itself is of secondary importance if a more expensive pump will save having to buy a larger solar array. Always look at the total lifetime cost of a system in terms of cost per gallon (see Chapter 11).

You will need to consult with the pump manufacturers to get the performance data. The retail supplier may also have this kind of information. Ask several retail suppliers which pump they recommend. If there is a well driller in your area who has experience with solar powered water systems, ask them. If your neighbor has a solar powered water system, talk to them.

Solar magazines, the Internet, solar expos, and the local library are good sources of current information (see Chapter 10).

Many suppliers sell complete systems, including the pumps, modules, and mounting structure. They will have advice for the size of system for your application. Buying a complete system from a supplier eliminates any finger pointing.

Before you contact your supplier, make sure you have the facts available. You will need to know the capacity of the water source, well static water depth and draw-down (change in depth) at different flow rates, diameter of casing, quality of water, sand content, height of water tank, distance of the water tank from the well, piping details, location and climate, water usage, what time of the year the water is needed, and anything else that the supplier may want to know.

Picking the right pump for your system will mean many years of reliable, trouble-free water supply. You may have to make adjustments to the number of solar modules used, but most pumps are tolerant to higher voltages and will respond to higher voltage with increased flow.

8
THE BASICS OF SYSTEM SIZING

In this chapter you will learn the basics of designing a solar system.

This chapter is merely an overview. Detailed information on system design, including software, is in the companion book *Solar Design*.

ENERGY AUDIT

The first step in solar system design is to do an energy audit. The exact ratio depends on many factors, but for every dollar spent in energy reduction, about 4 dollars can be saved in the cost of a solar electric system.

You can do an energy audit yourself or hire an expert. An energy audit is nothing more than looking around for ways to reduce energy usage. Some examples are:

- Replace incandescent bulbs with compact florescent bulbs. (Look for even lower energy LEDs for 120-VAC fixtures in the near future.)
- Low energy LEDs are available now for DC systems.
- Replace older appliances with energy efficient appliances (refrigerators, freezers, furnaces, heat pumps, air conditioning, washing machines, dishwashers, TVs, entertainment centers, computers, and fireplace inserts).
- Replace an electric clothes dryer with a gas clothes dryer or a clothes line.
- Use solar thermal to heat your water or buy an energy efficient water heater (Solar thermal is different than solar electric and is not covered in this book.)
- Install a blanket on your water heater and make sure the hot water pipes are insulated.
- Lower the temperature setting on your water heater.
- Cut back on hot water usage.

- Use a programmable thermostat to control your heating and air conditioning.
- Turn down the heating or turn up the air conditioning temperature, especially when you are away from your house.
- Eliminate (or unplug when not in use) phantom loads such as the small plug-in transformers that power such things as printers, and the instant-on appliances such as some TVs and stereos.
- If you use batteries, use an energy efficient charge controller and keep the batteries in good condition.
- If you have an inverter, use an efficient one and mount it in a cool location.
- Look for ways to reduce heat gain and loss in your home by adding additional insulation and reducing air leakage.
- Add skylights to reduce daytime lighting needs.

Many good sources of information on how to reduce energy use are on the Internet and in books. A professional energy audit can be very educational and worth the money, but common sense and being observant can also go a long way in saving energy.

If you are planning to hire an installer, do the energy audit first. The installer may not bring up this subject, after all they want to sell the largest system.

DETERMINE LOADS

In order to size the system, you will need to know what your energy needs are. If you have saved your utility bills, you can use them to estimate what your electrical consumption will be in the future. Take into account any reductions in electrical consumption by efficiency improvements, and any increases due to an expanding family or adding on to your house.

If you have not saved your utility bills, you may be able to contact the utility company or go online to get an accounting. You will need the electrical usage for the entire year. More years are better, as the electrical consumption can vary from year to year.

If this is a new house or an application where electrical consumption is unknown, such as a DC system in a cabin or boat, you will have to predict the energy needs. You will need to list the loads and the hours of operation. Estimate if exact values are unknown. Multiply the Watts times the hours to get the energy usage. Sum all the energy usage to get the total energy consumption per month in Watt-hours.

ACCOUNT FOR LOSSES

No piece of equipment is 100% efficient. A solar system has numerous losses associated with it (see Figure 66).

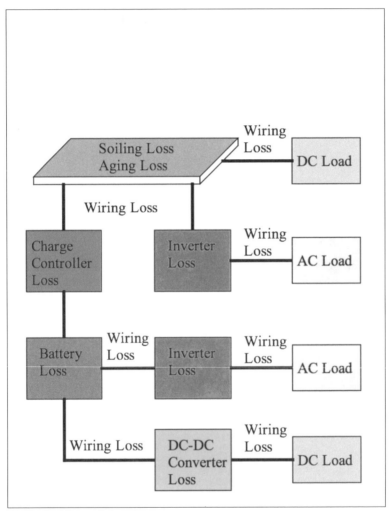

Figure 66 Solar Electric System Losses.

The losses in the inverter, charge controller, batteries, wiring, and other equipment increase the amount of energy needed. Changes in the module output over time take place due to soiling and aging.[77] These losses increase the needed size of the solar array. Accounting for losses is a reiterative process. The efficiencies of the different pieces of equipment are first estimated. After the system becomes more defined, and actual equipment is chosen, the values are adjusted and the size of the solar array is refined.

BATTERY SIZING AND CHARGING

If batteries are used, they are sized by the amount of days they need to supply energy, the daily energy need, the battery efficiency, and desired depth of discharge. If other sources of energy are available, such as an engine driven generator, this can often lower the overall system cost by reducing the battery cost <u>and</u> the generator cost. During a string of cloudy days, the battery bank doesn't have to supply energy for as long because the generator can be started. The generator doesn't have to run continuously because the battery bank is normally supplying the energy.

The amount of daily energy needed to charge the battery bank depends on the number of days available to accomplish this task. More days mean less daily energy needed and a smaller solar array. It is hard on the batteries to remain uncharged, so about three days is the maximum recommended time before fully recharging the battery bank.

CHOOSE EQUIPMENT

The equipment, such as the charge controller and inverter, are chosen to match the power, voltage, and current requirements. If there is a water system, the size of a water storage tank is determined in much the same way as the batteries: by the number of days of supply required and the daily water consumption. The water pump is chosen based on filling the

---

[77] Losses due to aging happen over a long time. Check with the module manufacturer to see what they warrant.

water tank in a certain amount of days and the solar is chosen based on how much power the pump requires.

SITE SURVEY

A site survey must be done to determine the orientation, tilt, and the amount of sunlight the proposed array will receive. The geographic location of the site affects the output due to the climate, altitude, and latitude.[78] Output is also determined based on whether the array is fixed or mounted on a tracker. If mounted on a roof, the orientation and tilt are given by the roof design. The system designer has to measure the roof size, angle, and orientation.

The amount of sunlight that array receives can be determined using tools like the Solar Pathfinder (see Figure 67).

Figure 67 Solar Pathfinder.[79]

---

[78] Sunlight is more intense at higher altitudes.
[79] http://www.solarpathfinder.com/

If the solar array is going to be mounted on a roof, the structure of the roof must be examined to determine if it will support the array. The roof may have to be repaired or reinforced. The condition of the roofing is important, too, as the array will have to be removed if the roofing has to be replaced in the future. Re-roof first. Chapter 12 has more on this subject.

Choose the Modules

The size of the solar array is determined based on all the factors discussed above: tilt, orientation, local climate, latitude, altitude, and daily energy consumption. Once the size of the array is known, the modules can be chosen. The module rated output times the number of modules add up to the required array output. Factors such as price, availability, aesthetics, and warranty, will affect which modules are purchased.

A computer program is the only accurate way to calculate the array size. The companion book, *Solar Design* includes just such a program, with a database of sites around the United States.

Wiring

The correct wiring sizes are chosen based on safety, acceptable losses, use, and cost. Safety always comes first, based on the National Electric Code (NEC).

Breakers or fuses and disconnects are required on solar electric systems. Again, safety comes first. The companion book, *Solar Design* includes a "Wiring Calculator" that helps in choosing wire size and breaker or fuse size. The wiring and over-current protection device must always be checked against the NEC and applicable local codes.

Equipment and wiring grounding is required for safety. Ground fault protection devices may also be needed for safety, as well as lightning arrestors.

The system designer is responsible for seeing that the system meets all the safety requirements and that it will function correctly. The designer must be familiar with the codes that govern the installation as well as local building department requirements.

Balance-of-System

The designer is responsible for specifying the balance-of-system. The mounting system is chosen based on the roof design, wind loading, seismic loading, snow loading, ventilation under the modules, ease of installation, and cost. The system designer must determine how the electrical lines are to be run, how the equipment is to be mounted, what outdoor enclosures are needed if any, water system components, etc. The task can be a large one depending on the size and complexity of the system.

Documentation

All the work needs to be documented so that parts can be ordered, permits obtained, and labor contracted. Documentation includes drawings, bill of materials, and calculations. The local building department will provide a list of required documentation. This is an important task, as it will make the project go smoothly or disjointedly, depending on the quality of the documentation. It can also affect cost, as a contractor is likely to add a cushion to a job with sketchy drawings or bill of materials. Good documentation can also eliminate costly mistakes.

9
FEDERAL AND STATE INCENTIVES

In this chapter you will learn what incentives are available to you at the Federal, State, and local levels. The chapter will cover various types of incentives, such as rebates, tax incentives, grants, loans, industry support, and production incentives. You will learn where to go to find out what is available to you. Since these incentives are changing all the time, you will not find a static list, which would be out of date by the time you read this. Instead you will learn how to find current information.

DSIRE

The best starting source for information about incentive programs is the "Database of State Incentives for Renewables and Efficiency" (DSIRE), brought to you by the North Carolina State University Solar Center.[80] The information is constantly updated and is all inclusive. Contacts are provided for further information.

FEDERAL INCENTIVES

Various kinds of Federal incentives were available at the writing of this book:

- Tax deductions for energy efficient commercial buildings
- Accelerated depreciation
- Tax exclusions for certain types of subsidies
- Tax credits for solar equipment
- Tax credits for solar equipment manufacturers, builders of energy efficient homes, and renewable electricity production
- Grants of funds for tribal energy programs
- Grants of funds for rural energy efficiency programs
- Federal low-interest loans and mortgages for clean energy, energy efficiency mortgages, and certain energy conservation bonds

---

[80] http://www.dsireusa.org/

- Loan guarantees for reduction of air pollutants and rural energy efficiency improvements
- Tax exemptions for energy conservation
- Tax credits for residential energy efficiency and renewable energy
- Tax credits for alternative fuels, alternative fuel vehicles, hybrid-electric vehicles, and electric vehicles

More or different incentives may be available by the time you read this. Consult with a tax attorney and direct the attorney to the DSIRE site. Unless you are familiar with writing grant requests, get help when applying for grants. Some incentives are only available to corporate entities. There are strict rules to follow, but you can save lots of taxes and even get funds for your project.

STATE INCENTIVES

Rebates

Rebates are funds provided to help defray the cost of a solar project or energy efficiency improvements. They are provided by state governments, local governments, and by local utilities. Rebates may be available where you live for grid-connected or stand-alone. If grid-connected, you must get approval from the local utility company, get a permit, and use a qualified contractor. Solar installers specialize in getting the paperwork done. You can sometimes sign over the rebate to the installer in lieu of part of the payment. As of this writing, rebates are available in 40 of the 50 states. California had 43 different rebates available. The amounts can vary widely.

Start the paperwork early. Use the DSIRE site to get the contact information for your state or local government offices and your utility company and contact them. Since energy conservation is a part of solar energy, start there. You may get a rebate for your new heat pump or get a check when the utility picks up your old, energy inefficient refrigerator that you would otherwise have to pay to haul to the dump. You can get rebates for additional insulation in your house. **Get paid for saving money.**

For solar systems, a rebate can make the difference between a 20-year payback and a 5-year payback.[81] Some rebates require that you use equipment from a qualified list and licensed contractors. Review the rebate requirements before getting started on the project. Contact the rebate provider.

See Figure 68 for a rebate example from the DSIRE site of a rebate available in Arizona.[82] This rebate requires that the equipment meet certain standards and that the equipment be installed by a licensed contractor on an approved list, but it does pay 50% of the cost up to 4 dollars per Watt. They also have a loan program available. The rebate provider retains the renewable energy credits, which are valuable as they can be bought, sold, and traded.[83]

LOANS

Loans are available from some state governments, local governments, and utilities. These are low-interest or no-interest loans for energy efficiency improvements and for renewable energy projects. The loans are available to private individuals, corporations, and public and nonprofit organizations. The loan rates vary as well as the term, usually 10 years or less. As of this writing, 37 of the 50 states provide loans. A sample loan for small businesses, from the DSIRE site, is shown in Figure 69.[84]

Energy efficiency improvements and solar projects are often funded by home improvement loans. You can write off the interest on these loans, just as you can do with any home mortgage.

---

[81] Payback time is the time it takes for the savings to equal the cost. If you save $5,000 per year and the cost is $25,000, the payback time is 5 years. More in Chapter 11.

[82] http://www.dsireusa.org/

[83] http://en.wikipedia.org/wiki/Renewable_Energy_Certificates

[84] http://www.dsireusa.org

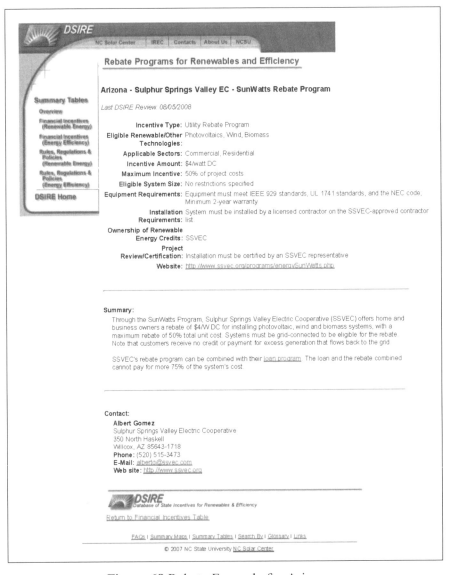

Figure 68 Rebate Example for Arizona, Sulphur Springs Valley Electric Cooperative.

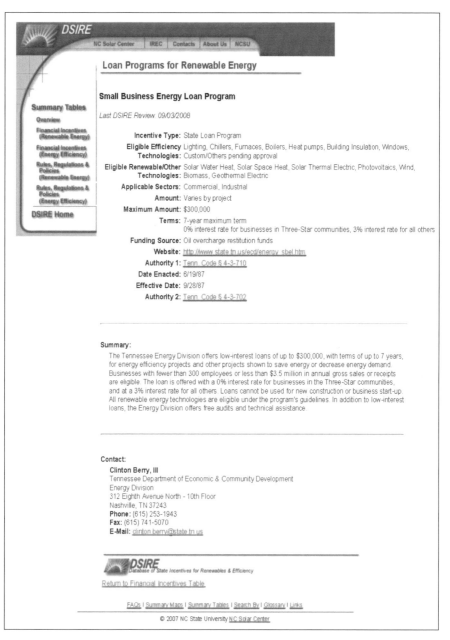

Figure 69 Loan Example, Tennessee Department of Economic & Community Development.

GRANTS

Grants are gifts of funds for worthwhile projects. State governments, local governments, utilities, and nonprofit organizations provide grants. Grants are primarily for small businesses and educational and government entities. Grants are like rebates in that they pay a portion of the cost of an energy efficiency or renewable energy project. As of this writing, 28 of the 50 states provide grants. A sample grant for schools, from the DSIRE site, is shown in Figure 70.[85]

TAX BREAKS

As of this writing, 47 of the 50 states provide some form of tax break for energy efficiency improvements or alternative energy projects. The tax breaks take the form of personal, corporate, sales, or property taxes. Consult with your tax attorney and show them the information available from the DSIRE site.

Personal tax breaks are applied to your state income tax. Corporate tax breaks are credits, deductions, and exemptions on corporate income taxes. Sales tax incentives are exemptions from sales tax for certain products and sales tax holidays for one or two days per year. Property tax breaks exempt taxes on the increased property value due to an energy efficient improvement or alternative energy source.

PRODUCTION INCENTIVES

Production incentives are cash payments for alternative energy production. The payments are usually based on the actual energy produced, kilowatt-hours, rather than the installed capacity, Watts.

Feed-in Tariff

A feed-in tariff is payment for alternative energy by the utility company at above market rates. As of this writing, feed-in tariffs were rare in the United States.

---

[85] http://www.dsireusa.org

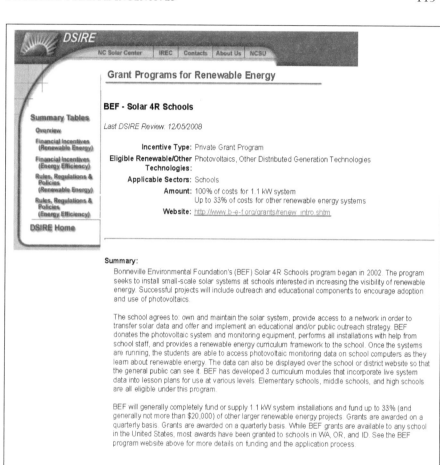

Figure 70 Grant Example, Bonneville Environmental Foundation.

BONDS

Bonds are a method for bonding agencies, such as state governments and utilities, to raise money for energy efficiency improvements or alternative energy equipment. The bonding agency sells the bonds to the public and then pays back the bonds through the energy savings.

WHAT TO DO IF YOUR STATE DOES NOT HAVE AN INCENTIVE PROGRAM

Get involved with the state governments, local governments, or local utilities. Attend public meetings about incentive proposals. If you are comfortable speaking in public, present a paper. One person can make a difference.

You can also help to elect representatives that support alternative energy. Be sure to research the background and voting record of the candidate. The support can take the form of a single vote, cash contributions, or working in the candidate's organization.

You can also get involved by doing a demonstration project in your town and writing about it. Newspapers are always looking for something new and different. Education goes a long way.

THE FUTURE

The price of traditional sources of energy goes up over the long run. When prices increase, look for more incentive programs for energy efficiency and renewable energy production.

Political winds change constantly. As they do, the support for solar energy waxes and wanes. If you see increased government funding for solar energy, such as rebates, take advantage of it because in a short time the funding will evaporate.

10
SOURCES

In this chapter you will learn where to find modules, inverters, mounting frames, and other solar equipment. The industry is changing all the time, so you will not find a static list of suppliers. Rather you will find the tools to search for yourself.

THE INTERNET

The Internet is one important tool to find sources. Even though there may be a good supplier in your home town, finding it may be easiest on the Internet. If you are planning to use an installer-contractor, again the Internet may be the best tool to find him or her.

HOW SEARCH ENGINES WORK

If you search on a large search site, like **www.google.com**, it will list results that have already been found. They have software that continuously searches the web and stores the results by keywords in their computers. That is how you can get millions of results in less than a second.

Smaller sites that target a specific area will list only the sites that have been entered by the website administrator. Businesses can request that they be listed.

The words and phrases that you search by determine what results are listed. The software that makes a website includes keywords. The site designer can carefully choose the keywords so that the site is likely to pop up first in the search results.[86] The first few sites shown are usually the only ones that get clicked on.

---

[86] The algorithm used by search engines to determine the order of the results is complicated and includes many factors besides keywords.

Websites also pay to be first. You might see a list of sponsored links or priority business listings. These sites have paid. The site may pay a small amount per click, perhaps one or two cents.

Searching on the Internet is an art. If you search using multiple words, such as discount solar modules, you will get all the results with the keywords "discount," "solar," "modules," "discount solar," "discount modules," "solar modules," and "discount solar modules" (3,450,000 results). The number of results can be overwhelming and you may not get what you are specifically looking for. However, if you use quotes, such as "discount solar modules," you will only get those sites that have the keyword phrase "discount solar modules" (6 results). This reduces the number of results and restricts the search to what you are specifically searching for. You can use a number of quoted phrases or words, such as "discount" and "solar modules" (18,400 results). You can actually be too specific so that no sites are found, in which case you can remove a few quotes or words.

If you don't find what you are looking for, try different variations, such as "discount solar" (23,500 results), "discount PV" (1,110 results), "affordable solar" (73,200 results), and "discount photovoltaic" (320 results). Try other words such as "low cost." Remember that you are trying to match up with the keywords that the site designer has put into the website.

COLLECTING BOOKMARKS

Several solar retailers are online. Once you find a site that is interesting, start a folder of bookmarks (favorites). Compare the prices for the same items. Check out the information available on the site. Is the physical address and the phone number of the site listed? If not, beware. Are there misspellings or incorrect grammar on the site? Try a phone call and ask a technical question or for a recommendation. Try an email. Ask how long they have been in business and what their annual sales are. Who owns the company? Are the employees listed on the site? How many employees? You can quickly gain some confidence that this is a company that you want to do business with, or not. Prices are important, but not everything. You want to be able to ask questions and get assistance. If you have problems, you want to be able to get them resolved.

ENERGY.SOURCEGUIDES.COM

The site that probably has the most complete list of renewable energy businesses worldwide is **http://energy.sourceguides.com/index.shtml**. At this writing, they had over 12,000 listings for renewable energy businesses and organizations and add more each month. Even so, they don't list everyone.

You can search by geographic location, product type, business type or by name. You can also search using keywords.

To use the site, you follow a decision tree. You continuously narrow your search by choosing different options, from broad to more specific, until you arrive at a list of companies. For instance, searching for "photovoltaic module retail businesses listed by business name," you would follow the decision tree below:

1. Source Guides
2. Renewable Energy
3. Renewable Energy Businesses
4. Renewable Energy Businesses in the World by Product Type
5. Solar Energy Businesses in the World
6. Solar Energy Businesses in the World by Type of Solar Energy Product
7. Photovoltaic Module Businesses in the World
8. Photovoltaic Module Businesses in the World by Business Type
9. Photovoltaic Module Retail Businesses in the World
10. Photovoltaic Module Retail Businesses in the World by Business Name

You will first see a list of Priority Business Listings, which are the sites that pay. If you click on the letters of the alphabet at the top of the page, you will get the remaining sites. Again, the ones listed at the top of the alphabetized lists have paid. You may not find all the companies because some are listed in a different category than the one you searched under, or they may not be listed at all. For instance, one retailer, not listed in the above category was found under Solar Water Pumping System Businesses in New Mexico, even though they supply photovoltaic modules.

If needed, use a major search engine, like Google, to find company contact information. Include specifics, such as the city and state. Remember to use quotes to narrow your search.

GOSOLARCALIFORNIA

The GoSolarCalifornia site lists equipment that is eligible for the California Solar Initiative (see **http://www.gosolarcalifornia.org/equipment/index.html**). This site is a joint effort of the California Energy Commission and the California Public Utilities Commission. The equipment can be used for any system. Other incentive programs reference this site for eligible equipment.

The site lists eligible solar modules, grid-connected inverters, and performance monitoring systems. It gives more realistic performance data, obtained from independent testing. It does not provide manufacturer's contact information. It has a search function for registered installers, contractors, and retailers in California. The site lists the Federal and California incentives available.

To use the GoSolarCalifornia site to find a module retailer, copy the module manufacturer information to a large search engine like **www.google.com**. Remember to use quotes.

SOLAR MAGAZINES

Solar magazines can be a great source. Examples are *Homepower Magazine*[87], *Solar Today Magazine*[88], *Renewable Energy Focus Magazine*[89], *Photon International*[90], *Home Energy Magazine*[91], and more. The magazines periodically list retailers or manufacturers. The ads in the magazines can be useful. If you are planning a solar project, get a subscription. Besides finding sources, the articles are a great way to learn about solar in general. Copies may be available at your local library.

Homepower Magazine

Homepower offers an online subscription at a reduced rate. It has a listing of installers in each issue. The articles are informative and the "Code Corner" provides expert advice on safety and code issues. Back issues are available.

Solar Today Magazine

Solar Today is published by the American Solar Energy Society, the US section of the International Solar Energy Society.[92] An online edition is available and the site includes older issues. The ads provide a source of information on products and services and the articles provide current solar news, technology, and new product information.

Renewable Energy Focus Magazine

Renewable Energy Focus is published by the International Solar Energy Society.[93] The magazine includes articles on international news and affairs, product news, and an international events calendar. The magazine has an

---

[87] http://www.homepower.com/
[88] http://www.solartoday-digital.org/
[89] http://www.renewableenergyfocus.com/
[90] http://www.photon-magazine.com/
[91] http://www.homeenergy.org/
[92] http://ases.org/
[93] http://www.ises.org/

online "Product Finder" feature where you can search for renewable energy products and services. At this writing the list of products and services was limited.

Photon International

Photon International includes articles on international solar news and technology. A sample edition of the print version is available. It has an online product search feature with a limited number of sources. You can also search for current and archived news articles.

Home Energy Magazine

Home Energy provides information on energy efficiency and alternative energy for homes. An online subscription is available.

BOOKS THAT LIST SOURCES

At the time of this writing, one book was available that included sources for solar energy products: *Real Goods Solar Living Source Book—Special 30th Anniversary Edition: Your Complete Guide to Renewable Energy Technologies and Sustainable Living (Real Goods Solar Living Sourcebook)* by John Schaeffer, September 1, 2007. Real Goods is an online retailer for alternative energy products. The book includes a catalog for their store. Look for the current addition. Ask at your local library. As mentioned earlier, written sources quickly become out of date.

SOLAR EXHIBITS AND FAIRS

Look for a solar exhibit or solar fair in your area. The American Solar Energy Society site includes information about the ASES National Solar Conference, which includes exhibits.[94] The ISES site includes information about the international ISES Solar World Conference.[95] The Institute of Electrical and Electronics Engineers sponsors an international conference, the IEEE Photovoltaic Specialists Conference.[96] The National Renewable Energy Laboratory site lists further events, such as the International Photovoltaic Science & Engineering Conference & Exhibition.[97]

Local solar tours and expos are sometimes available. Search under "solar expo," "solar fair," "solar tour," or other appropriate phrases. These can be a valuable source of information and a way to talk to a broad number of vendors in a short time. They are a good way to educate yourself.

COOLER PLANET

The Cooler Planet site is a search engine to find an installer in your area.[98] The installers pay for customers that are pre-screened by Cooler Planet. Tools and information about solar systems are available on the site.

NABCEP

The North American Board of Certified Energy Practitioners maintains a list of NABCEP qualified installers.[99] The list covers solar electric and solar thermal installers. The installers must pass certain requirements, complete an application, pass a written exam, and pay a fee.

---

[94] http://ases.org/

[95] http://www.ises.org/

[96] http://ieee.org/

[97] http://www.nrel.gov/solar/

[98] http://www.coolerplanet.com/

[99] http://www.nabcep.org/

SEIA

The Solar Energy Industries Association provides a list of members that permit their names to be shown.[100] The SEIA is an organization that advocates for the solar energy trades. The members agree to adhere to SEIA's Code of Ethics.

WORD OF MOUTH

If any of your neighbors have a solar system, ask them who did the job, what equipment they used, if they are satisfied with the system, and what they would do differently. Sometimes this can be the best way to find out what works and what doesn't work. If working with an installer, ask for references and follow up by calling the references. Ask to see their installation.

---

[100] http://www.seia.org/

11
CALCULATING COSTS

In this chapter you will learn how to calculate the cost of your system. The cost can be figured several ways from a very simple "System Cost" to "Payback Period." A spread sheet called "Costing Calculations" is provided with the companion book *Solar Design* to help you with the math.

TYPES OF COST CALCULATIONS

This chapter covers the following cost calculations:

- System Cost, dollars per Watt ($/W)
- Life-cycle Cost, dollars
- Cost of Electricity, cents per kilowatt-hour (¢/KW-h)
- Payback Period

DISCOUNTING

Discounting is an economic calculation that accounts for the higher value of present money when compared to future money.[101] The discount rate is determined based on interest rate and risk. The further in the future, the less value the money has.

When you install a solar electric system, you expect to get return on your investment in the form of electricity. The electricity has real value based on the rate that the utility company would charge or avoidance of other costs, such as keeping a generator running. Because the value of this electricity is provided to you in the future, it has less value than if supplied in the present.

If you do a discounted cash flow analysis, you would reduce the value of future electricity. The discount rate would be based on the interest rate you

---

[101] See http://en.wikipedia.org/wiki/Discounted_cash_flow

could get if you invested the money elsewhere and the risk. The risk is that the solar system will not deliver or that the rates from utility companies will drop.

However, there are other factors to consider. It is likely that the rates from the utility company will rise due to inflation and increased fuel and maintenance costs. Interest rates are variable depending on economic conditions. For these reasons, the value of future electricity is not discounted in the following cost calculations.

SYSTEM COST, $/W

The system cost in $/W is the cost divided by the rated output. Use a bill of materials or list of parts to calculate the total cost. For instance, if the total cost of the system is $80,000 and the rated DC power output is 9,360 Watts, the system cost is 8.55 $/W DC.

Use the rated power output, either DC or AC. You can also decide what to include in the cost. Most grid-connected systems will not include the cost of the house or land on which the solar system is placed. However, if you are planning a large utility scale system, you probably want to include the cost of the land. Taxes, financing, and other recurring costs are covered in the life-cycle cost and the cost of electricity.

You can also decide to subtract the cost of any rebates. For instance, if the rebate is 4 $/W, the final cost to the owner from the above example is 4.55 $/W.

When you are comparing two systems, the key is to be consistent. When communicating to others, include the assumptions. Is it DC or AC power? Is labor included? How about the upfront cost of financing, permits, attorney fees, structural roof reinforcement, etc.? Is profit included, or is just the basic equipment cost? If someone is quoting you a price, ask them to include a detailed breakout of the cost components and how the rated output is calculated.

LIFE-CYCLE COST, $

Life-cycle cost is the total cost over the life of the systems. Once you have calculated the system cost, you are on your way to calculating the life-cycle cost. Three important things have to be added: An estimate of the lifetime of the system; the recurring costs; any lifetime credits. Life-cycle cost is sometimes called "Cost of Ownership."

System Lifetime

It is difficult to estimate the expected lifetime of a system. Warranties are one measuring stick. Current warranties for solar modules are 25 years, with talk about going to 30 years. It is possible that they will last twice as long or longer. A reasonable lifetime is from 25 to 50 years.

Electronics have undergone tremendous improvements in recent decades. Inverters and charge controllers are more efficient and provide more beneficial features. The technology has matured to the point where equipment can be expected to last 20 years or more. The replacement may depend more on the desire to upgrade rather than outright failure. A reasonable lifetime for inverters, charge controllers, and other electronic equipment is from 15 to 25 years.

Lead-acid battery lifetime depends on how hard it is used, the quality and design of the battery, and how well it is maintained. If not abused, a good quality battery can be expected to last 20 years. All things equal, battery lifetime is longer if the battery is kept at room temperature and the depth of discharge is small. Consult with the battery manufacturer. A reasonable battery lifetime is 5 to 20 years.

The balance of the system, such as mounting frames, wires, etc., can be expected to last the life of the modules. One caveat is the quality of the installation. If module lead wires are not installed correctly they can abrade away in a short time causing an open circuit or a short. Ground wires, if not installed correctly, can corrode and lose contact in a very short time (see Chapter 12).

The lifetime of the system is the lifetime of the longest lasting component. The components with shorter lifetimes become recurring costs. For instance, if the lifetime of the system is chosen to be 30 years and the batteries last 10 years, three sets of batteries will be consumed in the lifetime of the system.

Recurring Costs

Recurring costs are any costs associated with the operation and maintenance of the system. These can include costs such as:

- Equipment replacement
- Repair
- Cleaning the modules
- Inspecting and testing
- Removing insects and rodents
- The cost of borrowing money
- Insurance
- Taxes
- Painting
- Website maintenance
- Accounting services
- Any other items associated with operation and maintenance

Credits

You will likely get a tax credit for your solar system. You may also get renewable energy production credits and may be able to write off the depreciation (see Chapter 9). Depending on your tax situation, this may be a substantial benefit.

Subtract any credits from the cost. The credits may be only good for a portion of the lifetime of the system. If doing the calculations by hand, multiply the annual credits by the years in effect. If using the "Costing Calculations" program supplied with the companion book, *Solar Design*, enter the actual years that the credits are good for in the program.

Life-Cycle Cost Example

The life-cycle cost is the initial cost of the system plus any recurring costs over the lifetime of the system such as equipment replacement and operation and maintenance costs. For instance, in the grid-connected example above, the lifetime was chosen to be 30 years and the costs are as follows:

- Initialed System Cost Before Rebate $80,000
- Initialed System Cost After Rebate $42,556
- Lifetime Equipment Replacement $5,000
- Lifetime Operation and Maintenance $29,000
- Lifetime Credits   ($20,340)

- Life-Cycle Cost Before Rebate $93,650[102]
- Life-Cycle Cost After Rebate $56,206[103]

Life-cycle cost is usually improved if the system lifetime is longer. At the end of the system lifetime, operation and maintenance costs can become excessive, causing the life-cycle cost to increase dramatically. At that point, it is usually time to scrap the system.

COST OF ELECTRICITY, ¢/KW-H

Once the life-cycle cost is known, it is only necessary to divide by the total energy generated during the lifetime of the system to obtain the cost of electricity. The total energy generated will depend on the location,

---

[102] 80,000 + 5,000 + 29,000 - 20,350 = 93,650
[103] 42,556 + 5,000 + 29,000 - 20,350 = 56,206

orientation, and whether the system includes a tracker. The software provided with the companion book, *Solar Design* can be used to calculate the energy production.

For the above grid-connected example, in Boulder, Colorado, with the array tilted 47° and pointed south, the output is 19,313 KW-hours per year. In the 30-year lifetime of the system, the energy production will be 579,390 KW-hours (30 x 19,313 = 579,390). Using the after rebate lifecycle cost of $56,206, the electricity cost will be 9.7 ¢/KW-hour (56,206 / 579,390 = 0.097 $/KW-hour = 9.7 ¢/KW-hour).

The same system mounted in Albuquerque, New Mexico, fixed at 41° tilt and pointed south, produces 22,274 KW-hours per year, or 668,220 KW-hours in the 30-year lifetime of the system. The electricity cost after rebate is 8.4 ¢/KW-hour.

PAYBACK PERIOD

The payback period is the time when your investment pays for itself. If you have a grid-connected system, this is the point where the savings in utility bills equals the cost. If you are replacing or augmenting a Diesel generator, this is the point where the savings from not using the generator equal the cost.

If you are calculating the payback period by hand, for each year, add all the costs and subtract all the credits and the avoided cost. Continue this for each year, adding in the result for the previous year. The year that the value goes negative, that is the payback period. The companion book, *Solar Design* includes software that does this calculation for you.

To calculate the payback period when replacing a generator, first calculate the electricity cost for the generator. Add in costs particular to the generator, such as fuel, oil, maintenance, replacement, engine overhaul, etc. When figuring the cost per kilowatt-hour, use the actual electricity used, not the nameplate output of the generator. The generator may be capable of 5 KW, but if only 1.2 KW is being used, use 1.2 KW when figuring the electricity cost. Initial costs for a generator are lower than a solar system, but life-cycle costs are usually higher.

The grid-connected example from above avoids paying electricity costs of 15.5 ¢/KW-hour.[104] With rebate, it has a payback period of 11.6 years in Boulder, Colorado. After that, the system is producing income by avoiding electricity costs. This assumes no increase in the cost of electricity. With any increase in the cost of electricity, the payback period is less.

The system can go back in the red if costly replacement parts are required or if a large repair bill is encountered. An example is damage by a large hailstorm. Insurance can offset these unforeseen costs.

INTANGIBLE CREDITS

Solar systems are installed for many other reasons besides the immediate financial benefits. These can be environmental credits, avoided cost of electricity distribution, public relations, or energy independence. Intangible credits are real and have dollar values, although they are harder to quantify.

One value of a nonpolluting energy system is that you are avoiding the cost of the pollution. To estimate the dollar effect of the pollution, include such things as medical costs, damage to the environment, etc. How many tons of carbon dioxide are not being produced because of your system, and what is the dollar value of that? What is the cost of an oil spill by a super tanker?

---

[104] The cost of electricity from the utility depends on the rate structure. Examine your electricity bill and call the utility company if you have questions. Solar electricity can displace the most expensive utility electricity if you use a lot of electricity and have a tiered-rate structure.

Consider the effect of not importing foreign oil and thereby removing support for those economic and political structures involved. Energy independence leaves us free.

If you are a utility, distributed solar energy reduces the cost and losses of the distribution system. It can eliminate the sizable cost of having to upgrade a transmission line. Solar energy meshes well with the load curve, reducing the need for peaking generators.[105] Solar energy also has very real value as renewable energy credits, which can be sold or traded.

---

[105] Peaking generators come online only when needed. They are more expensive to run than the base load generators which are running all the time.

12
INSTALLATION

In this chapter you will learn how to install a system properly. You will learn how to guard against mechanical wear and tear, how to avoid corrosion, how to avoid UV radiation damage, how to provide a proper mounting structure, how to provide for security, and how to keep insects and animals from damaging your system. The quality of the installation will directly affect the safety, lifetime, and performance of your system. The key word here is "quality workmanship."

MOUNTING STRUCTURE

Before you can install your modules, you have to provide a structure to mount them on. Residential installations are usually mounted on roofs.

Roofs

You expect your solar system to last 25 years or more. What about the roofing underneath the modules? The prudent approach is to re-roof before mounting modules, using a roofing material that will last as long as the modules. Some mounting systems require special sealing procedures for the attachments. If necessary, work with the roofer to assure that any attachments and roof penetrations are properly installed and sealed during the re-roofing.

Modules are normally required to be attached to the structure of the roof—the rafters that support the decking. The rafters are usually 2X6 or 2X8 or similar wood beams and the decking is usually plywood or pressboard (see Figure 71).

Figure 71 Typical Residential Roof Structure.[106]

The roofing is the membrane that provides the waterproofing. Commercial buildings will have metal rafters and metal decking.

Modules are available that are permitted to be laid on flat roofs without attachment. The modules are especially designed for this kind of installation (see Figure 72). Mounting frames for flat roofs are also available that use ballast instead of fasteners. Make sure that ballast is permitted by your local building department and that the roof structure can support the additional weight. If in doubt, consult with a local structural engineer who is familiar with the code requirements and structural requirements in your area.

---

[106] http://www.corybarnett.net/

INSTALLATION   133

Figure 72 Modules Designed to be Used
Without Attachment to the Roof.[107]

The roof is designed to support itself and wind load, seismic load, snow load, people walking on the roof, and rain load.[108,109] The loads are multiplied by a safety factor, which depends on how the structure is used. Dead loads are loads that do not move, such as the joists, sheathing, roofing, equipment, etc. Live loads are loads that vary; such as people walking on the roof, snow load, seismic load, and wind load. Wind load depends on how windy it is where you live and other factors such as the

---

[107] http://www.metrotuned.com/ Photo by Wilson Tai / Metrotuned
[108] http://www.awc.org/
[109] Detailed instructions on structural calculations are beyond the scope of this book.

height of the structure. Seismic load is caused by acceleration due to earthquakes and depends on how close you are to a fault zone and other factors such as the building design. Snow load depends on how much snow you get and the slope of your roof. If you have the plans for your house, the loads and safety factors used for structural calculations should be listed.

A typical mounting system is shown in Figure 73.

Figure 73 Roof Mounting System.[110]

The mounting frame and attachments are designed to distribute the load and tie the load to the structural members. Solar modules and the mounting structure weigh about 3 pounds per square foot (psf), depending on the

---

[110] http://www.conergy.us/

INSTALLATION   135

module and mounting system used. Live loads such as wind loads and seismic loads add to this dead load. The total load is usually within the safety factor used in the roof calculations. If you have questions about the frame and attachments, consult with the frame manufacturer. The frame manufacturer can usually provide structural calculations specific to your location, building, and roof structure.

Do not mount modules with inadequate support. Do not mount modules so that they overhang the roof at the eves, sides, or peak. The force of the wind is amplified at these edges and the modules may be blown off, possibly causing injury. Follow the instructions given by the mounting manufacturer. Do not use homemade hardware.

Inspect your roof structure. Climb up into your attic and look at the joists and sheathing or hire a house inspector. Is there rot or soft spots? Is there termite damage? Are the joists cracked or burned? The author has seen all of the above. Measure the joists, their spacing, and the length. Does the roof move easily or feel bouncy when you walk on it? Some roofs, especially commercial roofs, are designed just to the minimum requirement to save costs and may have to be reinforced. If you have doubts, hire a structural engineer to evaluate the roof structure.

If you have special requirements, such as a frame mounted above the roof to re-orient the modules, you will have to do structural calculations. Unless you are qualified to do these calculations, you will have to hire a licensed structural engineer or architect. The building department will want to see a wet stamp from the licensed engineer for you to get a permit. You can save money if you can do the calculations yourself and then get the licensed engineer to review them and stamp the plans.

Special modules are made that replace the roofing. They can be solar shingles or modules that interlock to form a rain-tight roof. If you are planning this type of roof, make sure that the roof structure is appropriate by consulting with the module manufacturer.

Ground-Mounted Frames

Ground-mounted frames have to satisfy all the structural requirements of roofs: wind load, snow load, seismic load, etc., but the safety factors are not as high because human life is not affected if the frame falls over during a wind storm. A ground-mounted frame is shown in Figure 74.

Figure 74 Ground-mounted Frame in Boulder, Colorado.[111]

The foundations have to be designed to fit the loads and the local soil type. Foundations are usually reinforced poured concrete, but can also be screw anchors or driven pilings. Poured concrete foundations use a cardboard tube to form the part that is above ground level (see Figure 74). To design a proper foundation, you will need to know your soil type, usually determined by a soils engineer who will dig a hole. You need to calculate the loads from all directions and then design the foundation for the

---

[111] http://freedomsolarpower.googlepages.com/

INSTALLATION

allowable side load, uplift, and down load of the soil. If you do not know how to do this kind of calculation, consult with a structural engineer.

If you are building the frame from scratch, you will have to do the structural calculations or get someone to help you. If you buy a ready-made frame, the manufacturer can provide the calculations. They may supply a frame specifically designed for the loads in your area, such as wind load and seismic load.

Snow

If snow is heavy or frequent in your area, mount the modules at an angle that will shed the snow. 45° is usually enough. Do not have a lip or other restriction at the bottom of the modules that will keep the snow from sliding off. Raise the bottom of the array several feet above the level of average snowfall (see Figure 75).

Figure 75 Village Power Array Designed for Snow.[112]

---

[112] Courtesy DOE/NREL, PV/Diesel hybrid system for a remote village power system in Lime Village, Alaska, Northern Power Systems

THERMAL ISSUES

Heat reduces the output of charge controllers, inverters, and modules. The higher the temperature is, the shorter the life of the electronics will be. As stated earlier, but worth mentioning again, keep the electronics in the shade. Keep all parts in the shade, including any heat sinks.

Allow airflow on the back side of the modules. Heat is transferred from the front side and the back side. Blocking airflow on the back side will raise the module temperature 20°F or more. Install the modules at least 3 inches above the roof, more if possible.

Provide airflow for the electronics too. Inverters and charge controllers come with instructions on how much space to allow around them. If you have to mount them in a restricted space, provide a fan to keep the air moving.

INSTALLING BATTERIES

Batteries should be kept at room temperature (50°F to 80°F). Their lifetime is shortened at higher temperatures and the capacity is diminished at lower temperatures. Install batteries in insulated spaces. If you have a basement that keeps a uniform temperature year round, install the batteries there. If using a battery house, build it using concrete block to provide thermal capacity and insulate it.[113] If possible, build the battery house into the ground. Use large roof overhangs on the sunny side. If necessary, provide a heating source in the winter and a cooling fan or small air conditioner in the summer. Keep batteries off cold floors and provide air circulation.

---

[113] Thermal capacity, or heat capacity, is the ability of materials to store heat, like a flywheel stores energy. Water, brick, concrete, and similar materials have high heat capacity. The ground has excellent thermal capacity. A cave will maintain a nearly constant temperature year round.

Mechanical Wear and Tear

In most cases, wiring from module to module and to the combiner box is flexible cable with connectors, usually supplied by the manufacturer. The connectors are required to be waterproof and not capable of being pulled apart easily. Newer connectors include a snap together feature that prevents the connectors from being pulled apart.

The flexible wire is not static but will move and vibrate in the wind. If the wire is stretched too tightly or run over sharp edges or the abrasive roofing, it will eventually abrade in two. The flexible wiring is not designed to be self-supporting. The proper installation is to use stainless steel clips to support the cable every two feet or more, or use electrical conduit or wiring trays. Do not allow the cables to rest on sharp edges, such as the corner of the mounting frames. Gather extra cable and bundle with stainless fasteners. Installing new connectors on module wires requires special training and special tools, provided by the connector manufacturer.

Be careful when working with the modules so that you do not damage the wires. Do not lay the modules down on the wires. Do not lift the modules by the wires. Do not put kinks in the wires but use gentle bends.

When installing crimp connectors, it is imperative that you use the proper equipment. The common crimping tools available in hardware and electronic stores are not adequate. Use ratcheting, compound-action crimpers (see Figure 76). For larger connectors, use a heavy-duty crimper (see Figure 77). Improper crimping will not provide a tight connection. The connectors will eventually come loose or overheat due to resistance and can cause fires.

Figure 76 Ratcheting Compound Action Crimpers.[114]

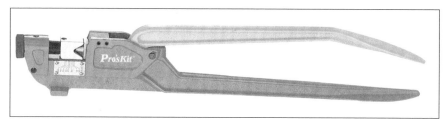

Figure 77 Heavy Duty Crimping Tool.[115]

Test the connections by pulling. Use a reasonable amount of force; don't overdo it. Module and connector manufacturers have a specification for how much pull to apply. The author has seen wires that easily pull out of junction boxes on the back of modules and the fires that resulted when these wires came loose. The module manufacturer is required to use a strain relief at the junction box. Check to make sure that the strain relief is tight.

Wires that are connected by screws or bolts, like on breakers, have a way of working loose too. When installing wires using a screw or bolt, torque the

---

[114] http://www.idealindustries.com/
[115] http://www.proskit.com/

INSTALLATION   141

screw properly.[116] After the installation is complete and before starting up the system, go back and recheck the torque. After the system has been in operation for a time, less than a year, go back and recheck the torque again.

Use a locking device on all fasteners. Use a locking washer, a metallic locking nut, or a locking adhesive. Use Nylon locking nuts and screw only where they will not be exposed to sunlight and only where the fasteners are installed only once. If the equipment is disassembled and re-installed, use a new Nylon locking nut or screw.

Drilling Holes In Module Frames

If you must drill a hole in the flange of an aluminum module frame, first make sure that it will not void the warranty. Second, protect the back side of the glass part of the module so that the drill will not hit the glass and break the module. Use a board or similar object under the flange.

CORROSION

Use an electronic grease, spray, or plastic coating to protect the terminals, especially if you are mounting your solar modules in wet environments, such as boats, seaside residences, marine navigation equipment, or any humid locations. You can find this material at marine supply stores or electronic supply stores. Assure that the connectors, junction boxes, and combiner boxes are rain tight. Do not misuse watertight wire feed-throughs (gland connectors), for instance by using multiple wires in a feed-through designed for one.

If you are in a wet environment, especially on an ocean-going boat, use marine grade wires. Marine grade wires have a tin coating that protects the copper wire from corrosion. Without the tin, the corrosion can start at the

---

[116] Use the torque specification that came with the equipment. If no specification available see http://www.torqwrench.com/ or other torque specification table.

end and work its way up the wire, making replacement of the whole wire necessary. You can obtain marine grade wires at marine supply stores or online. Beware that the marine grade wire may not have the sunlight resistance of sunlight resistant wires and may not have a high enough temperature rating. Consult with the wire manufacturer.

Use stainless steel hardware, such as sheet metal screws, nuts, washers, and bolts. You can find stainless steel hardware at marine supply stores or order online. If you are using aluminum, such as the module frames, make sure that it is hard anodized. Soft anodizing is much thinner and will not last as long. This is especially important if you are in a marine or wet environment. Stainless and aluminum come in different grades, some more corrosion resistant than others. For instance, 316 stainless is more corrosion resistant than other grades of stainless steel. One test is using a magnet. If the magnet sticks, it is a lower grade of stainless or not stainless. Even 316 stainless will develop surface rust on a boat in the tropics. Copper will corrode aluminum and should not be left in contact with any aluminum part.

If you are using galvanized steel hardware, use hot dipped galvanized. The electroplated galvanized hardware has a thin coating and will not last as long. You can tell the hot dipped coating by the rough texture. The zinc coating is sacrificial and thickness determines the lifetime. Use hot dipped fasteners on hot dipped hardware.

When using dissimilar materials in a wet environment, such as stainless steel fasteners in aluminum module frames, it is important to prevent an electrolysis reaction. The stainless will cause the aluminum to rapidly corrode. Use a lanolin containing grease, available at a marine supply store. Coat the fastener with the lanolin grease before installing it. The lanolin grease will also prevent water from standing in contact with the stainless, which will cause crack corrosion.

If you are building a wood frame for your modules, use treated lumber to assure a long lifetime. Follow safety procedures when cutting this poison-containing wood and do not burn the scraps, which are hazardous waste. The scraps do not go to the landfill and are not to be used by children as toys. Use a dust mask and eye protection.

Special hardware is available to attach grounding wires (copper) to module frames (aluminum) (see Figure 78).

Figure 78 Proper Module Grounding Hardware.

Proper grounding hardware will prevent corrosion. The copper wire attached to the aluminum frame with stainless hardware is asking for trouble and the modules may become ungrounded in a few years. The proper grounding lug is available from your solar equipment supplier. To obtain good electrical contact, sand or scrape the anodizing away from the frame where the grounding lug is attached.

Mounting frame manufacturers are coming up with grounding systems that eliminate the need to ground individual modules. If you decide to use this system, call the manufacturer and ask for documentation that shows that the hardware meets the National Electric Code requirements.

One last word about corrosion: do not allow standing water. Assure that enclosures are rain proof and have a drain hole. A loosely fitting stainless cotter key in a drain hole will keep it clear. A sealed box will eventually contain water due to the permeability of most materials. A better solution is to allow airflow without allowing rain to enter.

UV RADIATION DAMAGE

Ultraviolet (UV) radiation is the high-energy radiation coming from the sun. Most gets filtered by the upper layers of the atmosphere, but enough arrives at the surface to damage certain plastics and human skin. The UV radiation causes chemical changes in plastics and certain rubber components that weaken the material and eventually cause it to break.

Some plastics and rubbers are naturally more resistant to UV radiation. Others have components added that absorb the UV or keep it from getting beyond the surface. Black tie wraps are an example. They are the same material as the white tie wraps, but have a black pigment added that prevents the UV from entering the bulk of the material, and will last several times longer than the white tie wraps. Sunscreen is an example of a material designed to absorb UV before getting to your skin. No additives are 100% effective and most plastics will fail in sunlight given enough time.

Use materials that are UV resistant, such as glass, metals, ceramics, concrete, etc. Use sunlight resistant wires. If you have to use a plastic or rubber component, make sure that the material is UV resistant. Usually black materials are more UV resistant. Silicon rubber lasts a long time in sunlight, especially the black or colored variety. Acrylic without additives is sunlight resistant.

If possible, keep plastics and other organic materials in the shade. Keep the wires and connectors behind the modules and frames.

ANIMALS AND INSECTS

Rodents have to chew to keep their teeth trimmed. Somehow, they love to chew on electrical wires. They usually do not chew through the conductors, but the insulation is breached, causing a shock hazard and a fire hazard (see Figure 79). If rodents are a problem in your area, you will have to run the wires in conduit.

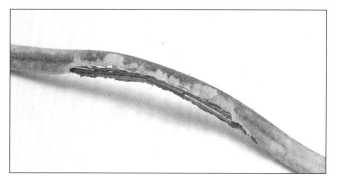

Figure 79 Rodent Damaged Wires.[117]

Do not make a home for rodents or insects. Mice can get into some very small holes. Do not provide them with the chance. Use chew-resistant enclosures, such as metal, or use a wire screen to keep them out. If they move in, hire an exterminator to evict them.

Conduit

Some code districts require conduit, especially if children or pets can come in contact with the modules. Kids can get on roofs or climb on module mounting frames. Make sure children do not play around any electrical equipment.

---

[117] http://www.southernhighlandspestcontrol.com.au

If you are planning to run conduit, make sure that the junction boxes on the backs of the modules allow conduit. Another option is to run the wires in wire trays (see Figure 80).

Figure 80 Metal Wire Tray Used To Protect Wire.[118]

Insects can also move into enclosures with small holes. Seal all enclosures or use an insect screen. If you find a wasp nest under your module, you can purchase a spray that will kill them. Read and follow the directions. Clean any residue from the surface of the module.

Provide a fence to keep animals out. If you are planning a water pumping system, the fence might be to keep cattle or horses from damaging the solar system. If you are planning an installation in Africa, you may have to

---

[118] http://www.mpandc.com/

keep monkeys or elephants out (see Figure 81). You can also mount the modules on poles or high legs to keep them safe.

Figure 81 An Orangutan-Proof Fence at Camp Leakey, Borneo.[119]

Birds will land on modules and deposit a shadow-producing material. This is especially true near oceans. To prevent this, you can install chicken wire an inch or so above the modules or mount the modules at an angle that prevents birds from landing. Strips of stainless steel spikes are available from marine supply houses which also supply such things as scare owls, wind-driven sweeps, and plastic strips. Some people have had luck stringing scrap CDs. Ask around to find what works best in your area.

---

[119] http://www.4015.com/orangutan/. The shadowing is a regrettable consequence of having to keep the orangutans from using the array as a jungle gym.

SECURITY

It is an unfortunate fact of life that certain people will steal your modules or damage them by vandalism. If this is a problem in your area, you will have to take precautions.

Theft can be prevented in several ways. Keep your modules behind a locked fence (also good for safety). Bolt them or lock them to the frame in a way that makes it impossible to remove them. Keep them out of sight by installing them on the back side of your property or building or a long way from the road. Make them inaccessible by putting them on your roof or a high pole. Install a security system. The best approach for you will depend on your situation. Make sure you have insurance that covers theft.

Protect your modules from vandalism in much the same way as you would protect them from theft. Install them out of sight or a long way from the road. Put them on your roof. Use a security system. Post your property and use fences that discourage trespass, such as electric fences. Make sure you have insurance that covers vandalism.

QUALITY WORKMANSHIP

Doing the job over again is expensive. Doing it right the first time saves money and assures that the system will safely perform for 25 years or more. To learn more about how to correctly install systems, talk to suppliers and installers or search online.[120] Visit solar installations in your area. Make notes of what you like and don't like. A quality installation is something to take pride in and will pay you back many times over through the years.

---

[120] Excellent examples of poor and good installations are at http://www.mpandc.com/

13
INSTRUMENTATION AND TESTING

In this chapter you will learn about the many forms of instrumentation that you can use to measure the performance of your system and trouble shoot lack of performance. The chapter ranges from simple instrumentation, such as multimeters, to more complex instrumentation such as data loggers.

THE NEED FOR INSTRUMENTATION

You need to know how your system is doing. Otherwise, you are blind to any changes or problems that may crop up. You will need to do an initial test of your system, known as a benchmark test, so that you can check the design and compare performance later on.

How much instrumentation you use is really a matter of personal choice and need. You may be happy with the information that the inverter or charge controller provides, or a voltmeter on the battery bank. Or, you may want to keep constant tabs on the performance with an IV curve tracer and data logger.

You cannot look at a wire or solar module and tell how much voltage or current is flowing. For that you need a meter. You also need a meter for a precise measurement of the amount of sunlight falling on your array. You are already familiar with the meter used to measure temperature, a thermometer, but you are going to need one that attaches to the back of the modules.

The Basics

At a minimum you will need a digital multimeter. Get one that can read several hundred Volts AC and DC, at least 10 DC Amps, and that can measure resistance from 0 to several megohms. Higher priced instruments are generally more accurate (see Figure 82). The digital multimeter has a response time of about a second. Most voltages and currents in solar applications are fairly constant, so the long response time does not matter.

Figure 82 Digital Multimeter.[121]

If you are not familiar with the digital multimeter, you will find it a valuable tool and discover that you are using it all the time. You will be using it not only on the solar system, but also to check your car and household appliances.

---

[121] http://www.fluke.com/

Instrumentation and Testing

The digital multimeter has a minimum of three functions: voltage, resistance, and current. Some multimeters are self-ranging, meaning that they will change the range of the reading in response to the value. You will have to change the leads on the meter to read current; if you forget to change it back to read voltage, you can ruin the meter. To measure voltage, touch the two probes to the locations where you want to measure voltage difference. Measure resistance in the same way, but turn off the power first. Measure current by inserting the meter in series with the circuit. This means that you will have to disconnect the wire and connect it to the lead from the meter.

Insulated clips are handy for current measurements and for multiple voltage measurements. For probing a fault, you can clip one lead on the ground or grounded conductor, and probe the other parts of the circuit with the other lead.

Remember that you may be dealing with a dangerous voltage. Use the multimeter with care. Read and follow the instructions. Use one hand if possible and hold the other hand in back of you so that you don't inadvertently become a path for the current. Do not disconnect the plugs on the meter while the probes are connected to a circuit. Always test for voltage before working on a circuit. Do not assume that because the switch is off there is no voltage; there could be fault somewhere. See Chapter 15 for more on safety.

You can use the multimeter to measure the open circuit voltage (Voc) and the short circuit current (Isc) of a module. These two points can tell you a lot about the module, especially if you know the sunlight intensity (see below). Obviously, no voltage or no current would indicate a broken module, and you don't need to know the light intensity to tell that. Use the multimeter in the voltage mode to measure the Voc. Use it in the current mode to measure the Isc. In both measurements, connect the probes to the module leads. Make sure that the module does not produce more current than the multimeter can measure. In the voltage mode, there is a tiny amount of current flowing, but not enough to affect the Voc reading unless you are measuring one cell. Details about correcting these measurements to standard conditions appear later on in this chapter.

If wired in an array, you will need to disconnect the module connectors to measure each module. Do not disconnect the connectors while the current is flowing or an arc in the connector can cause damage. Use the disconnect switch or breaker to disconnect the current first. The voltage may still be high, so be careful. Disconnect the first module from the grounded conductor first (see Figure 83).

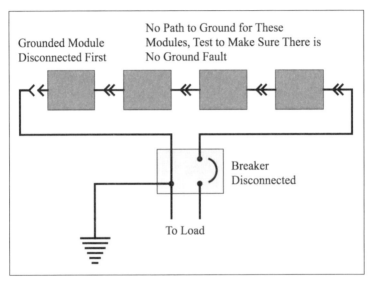

Figure 83 Disconnecting Modules in Series.

With the grounded conductor and the main disconnect both disconnected, there should be no path to ground.[122] Measure the voltage at the module side of the disconnected breaker in reference to ground. The voltage should measure zero. If it is not zero, there is ground fault within the modules or wiring. Disconnect the rest of the modules starting from the grounded module. If there is a ground fault, test for voltage at the breaker after each module is disconnected. When the voltage goes to zero, the ground fault is in the last module disconnected or the wiring for that module.

---

[122] This is similar to birds perched on a high voltage wire. They do not get shocked because there is no path to ground.

You can make up special leads for measuring the Voc and Isc. Use a DC rated disconnect in the lead so that you will not be disconnecting the connectors under load. Use the same connectors as the module uses. This will allow quick connection. Remember to disconnect the module before switching the plugs on the meter between voltage and current. If you are going to do a lot of connecting and disconnecting, use special connectors available from the manufacturer that are designed for lots of connections and disconnections. The standard connectors are only designed for about 10 or 20 disconnects.

Special ammeters have clamp-on loops to measure current (see Figure 84).

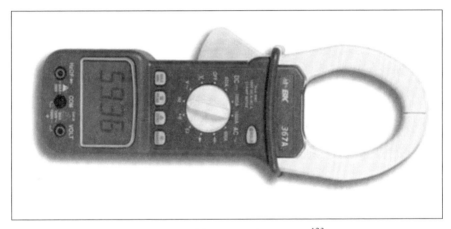

Figure 84 Clamp-on Ammeter.[123]

Clamp-on ammeters work by measuring the magnetic field around the wire which is proportional to the current. You need access to one of the current-carrying wires. You can make up a special extension cord with separate wires for measuring the current of AC appliances. AC current is easy to measure with a clamp-on ammeter. Because DC current is more difficult you will need a special AC/DC clamp-on ammeter. The stray magnetic fields that are all around us affect the accuracy of DC clamp-on ammeters. You will have to zero the ammeter close to the wire before clamping on to

---

[123] http://www.bkprecision.com/

take the measurement. Clamp-on meters are not as accurate as other means of measuring current.

Currents larger than the capability of the digital multimeter can be measured using a shunt resistor (see Figure 85).

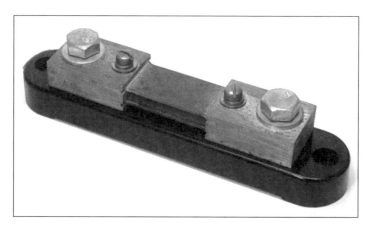

Figure 85 Current Shunt Resistor, 50 mV at 200 Amps.

A shunt resistor is a precision resistor that is inserted into the circuit where you want to measure the current. The voltage across the shunt resistor is proportional to the current. The voltage drop is small so that it will not adversely affect the circuit, such as 0.05 Volts (50 millivolts (mV)) for 200 Amps. The exact ratio depends on the size of the shunt resistor. The large screws or bolts are for the wire lugs that carry current and the smaller screws are to attach wires lugs for the multimeter or other voltmeter. You will measure the voltage and apply a calibration factor to determine the current. In this case, every 10 mV equals 40 Amps (200 / 50 X 10 = 40).

MEASURING THE AMOUNT OF SUNLIGHT

A solar system's current is directly proportional to the amount of sunlight, in solar circles called irradiance.[124] Twice the irradiance gives twice the current. If you don't measure the irradiance, you will not know if the

---

[124] Irradiance is the rate of solar radiation in Watts per square meter. Isolation is the irradiance over time in Watts per square meter per hour.

INSTRUMENTATION AND TESTING   155

system output is low because of reduced irradiance or because there is a problem.

You will need to measure the global irradiance at the plane of the array. Global means the light coming from all parts of the sky and plane of array means that you mount the instrument next to the array and pointing the same direction as the modules.

The most accurate way to measure global irradiance is to use a precision spectral pynanometer (see Figure 86).[125] The pynanometer from Eppley Laboratories are used by large testing facilities and large solar installations. They are the most accurate, but expensive.

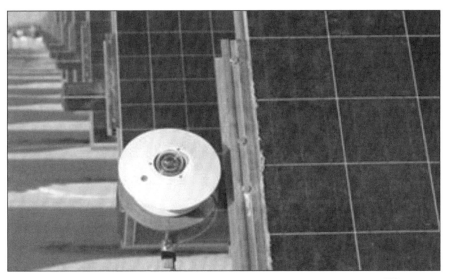

Figure 86 Plane of Array Precision Spectral Pynanometer.[126]

---

[125] http://www.eppleylab.com/
[126] Courtesy NREL/DOE, Steve Wilcox

A less expensive pynanometer is available which has a small solar cell inside (see Figure 87). Since it uses a solar cell, it is sensitive to the light spectrum.

Figure 87 Solar Cell Based Pynanometer.[127]

MEASURING TEMPERATURE

You will need to know the junction temperature of the cells to correct your data to standard conditions. The temperature on the back of the module is very close to the temperature of the junction. The temperature of the module is not uniform but cooler at the edge were the wind is coming from or the bottom if there is no wind. You can measure the temperature one foot in from two diagonal corners and average them or measure the temperature at the center of the module. The same goes for panels and arrays; the temperatures of the modules in an array are not uniform. Measure the temperature at the center of the two diagonal modules or one module in the center of the array.

An accurate way to measure the temperature is to tape a thermocouple to the back of the module. A thermocouple is the junction between two wires made from different metals. Two of these junctions at different temperatures will produce a small voltage. One junction is kept at a known temperature, such as in a jar of ice water, and the other is placed where you want to read the temperature. Modern instruments have an electronic reference junction built into them so that you don't have to carry around a

---

[127] http://www.apogee-inst.com/

bucket of ice (see Figure 88). Some multimeters have a thermocouple reading function, which can be very handy.

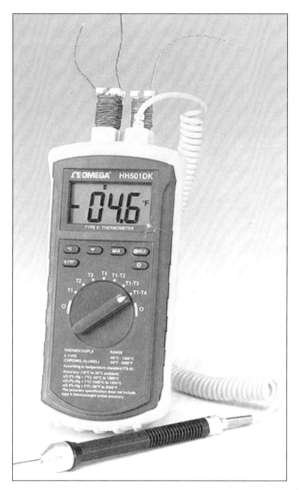

Figure 88 Thermocouple and Thermocouple Reader.[128]

Another way to measure temperature is to use an infrared (IR) thermometer. These instruments are very handy for checking temperature, but not as accurate as a thermocouple. You just point and click. The field of view is a cone, so the farther away you are from the object, the greater

---

[128] http://www.omega.com/

the area you are measuring. Some have laser sights to help you aim at the spot you want to measure (see Figure 89).

Figure 89 Infrared Thermometer.[129]

Other means to measure temperature are thermistors and resistance temperature detectors (RTDs), which are more accurate than a thermocouple if used correctly. You can also use temperature paint and stick-on devices that change color with temperature, but these are less accurate and harder to use.

---

[129] http://www.reedinstruments.com/

MEASURING THE IV CURVE

Measuring Voc and Isc is a start, but will not give you a complete picture. To really see how a module or array is performing, you need a complete IV curve (see Figure 90). If the module or array has developed a problem, you are likely to see a change in the shape of the IV curve. The shape is quantified by the fill factor (Ff) which is the voltage times current at peak power divided by the Voc times Isc (Ff = (Vmp X Imp) / (Voc X Isc)). See the companion book *Solar Design*, Chapter 9, for a more detailed discussion of IV curves.

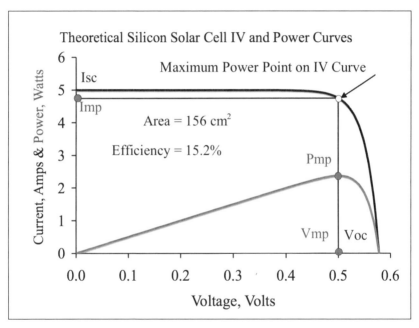

Figure 90 IV and Power Curves of a Solar Cell.

The IV curve of a module can be measured using two multimeters and a variable resistor or resistor bank. This method is not recommended for arrays because of the power involved. One multimeter is used in a voltage mode and one is used in a current mode (see Figure 91). Make sure the variable resistor or resistors can dissipate twice the power generated by the module and has a minimum resistance that will load up the module.[130]

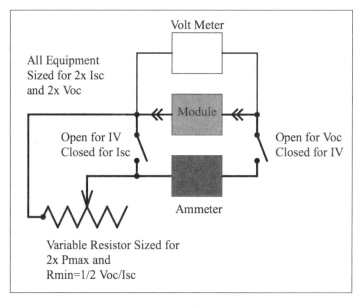

Figure 91 Setup For IV Curve Measurement.

Take readings of voltage and current at several points, multiplying them together to get the power. Search for the maximum power point. Do this with constant amount of irradiance. This is actually the hard way to measure an IV curve. Using an IV curve tracer, discussed below, is much easier.

You can replace the variable resistor with a programmable power supply that will accept power as well as generate it. You can also replace the multimeters with meters that will interface with your computer. Then you can write a program that will take the IV curve for you, do the corrections, and print out the results. You will then have the same equipment that is used in module factories.

---

[130] You will not be able to measure true Isc or true Voc with a resistor in series with the module. Use a variable resistor with a minimum resistance of half of the Vmp divided by the Imp (R ≤ 1/2 X Vmp/Imp). It must also take twice the rated power, (2 X Pmax). This is a big variable resistor.

An IV curve tracer is a machine that automatically takes an IV curve (see Figure 92). You will need to hook up the wires to the array, hook up a pynanometer, hook up one or two thermocouples to the back of the module, and hook up a laptop computer. These instruments are accurate and can measure large arrays, but are expensive. You can sometimes rent them from your local solar installer or solar equipment supplier.

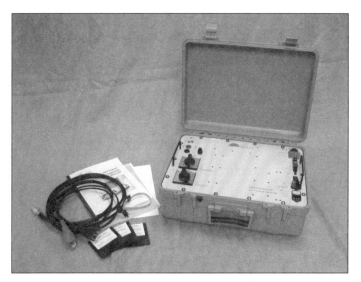

Figure 92 Curve Tracer.[131]

CORRECTING TO STANDARD CONDITIONS

Standard test conditions (STC) are 1,000 Watts per square meter (W/m^2) and 25°C cell junction temperature. Results are reported at standard conditions so that different modules can be compared and so that you can compare your system performance over time and know if there is a problem.

As an approximation, voltage varies with temperature and current varies with irradiance.[132] Each current value on the IV curve will be multiplied by the ratio of 1,000 divided by the measured irradiance. It helps the accuracy if the measured irradiance is close to 1,000. For instance, if you measured

---

[131] http://www.zianet.com/daystar/

[132] Current actually increases a small amount with higher temperature.

an Isc of 4.85 Amps and a Imp of 4.45 Amps at 875 W/m^2, the corrected Isc would be 5.54 Amps (4.85 X 1,000 / 875 = 5.54) and the corrected Imp would be 5.09 Amps (4.45 X 1,000 / 875 = 5.09).

To correct the voltage to standard conditions, each voltage on the IV curve is multiplied by a voltage correction factor. To determine the voltage correction factor, first correct the Voc by using the Voc temperature correction factor supplied by manufacturer. For example, if the measured temperature on the back of the module was 38°C, the measured Voc was 14.5 Volts, and the Voc correction factor was -100 mV/°C (-0.100 V/°C), the corrected Voc is 15.8 Volts (14.5 - 0.100 x (25 - 38) = 15.8).[133]

The voltage correction factor is then the corrected Voc divided by the measured Voc, or 1.090 (15.8 / 14.5 = 1.090). Each voltage is then multiplied by this voltage correction factor. For instance, if the measured Vmp was 13.8 Volts, the corrected Vmp is 15.0 Volts (1.090 x 13.8 = 15.0).

The companion book, *Solar Design* includes software to do the calculations for you. Also, if you are using a curve tracer, the calculations are done for you. For the best accuracy try to take data when the junction temperature is close to 25°C, which in most cases means winter.

KEEPING TRACK OF ENERGY PRODUCTION

More and more inverters and charge controllers are provided with built-in data loggers that will keep track of the system production. The information is displayed on the equipment and/or on a separate remote display. Typical information displayed on charge controllers is:

- State of charge for the battery bank as a percentage or bar graph
- Current and voltage from the solar array
- Charge and discharge current from the battery bank
- Battery bank voltage
- Charge mode (equalization, bulk, absorption, or float)

---

[133] mV stands for millivolt, which is 1/1,000th of a Volt

- Current from a separate generator or utility company source if available

Depending on the charge controller, a record of this data may be available. A computer interface may also be available that will allow you to download the data.

Typical information displayed on an inverter is:

- Array current, voltage, and power
- Grid current, voltage, power, and frequency
- If used, generator current, voltage and power
- Daily peak capacity
- Date and time
- Daily yield
- Annual production and annual net energy delivered to the grid
- Total production and total net energy delivered to the grid
- Daily, annual, and total hours of energy production
- Avoided carbon dioxide generation
- Battery voltage, state of charge, and energy in/out

As with the charge controllers, a record of this data may be available from the inverter.

If you have a unique situation or don't have a charge controller or inverter with the built-in data logger, you can buy a separate data logger. A data logger is nothing more than a set of computer-controlled switches that connect different measuring lines to a voltmeter. The data is then fed to a computer or stored in the data logger for later retrieval. The voltmeter is usually built into the data logger.

Data loggers are available that plug into the motherboard on your computer or are separate units. Handheld data loggers are also available. You run software on your computer that controls the data logger and stores the data for later display. Data loggers can be configured to display almost any kind of information, including irradiance, wind speed and direction, and the temperature of each individual module. Research and testing facilities use lots of data loggers.

With the right instrumentation, you can connect to your inverter or charge controller over the Internet to see how the solar system is doing. Satellite systems are even available for monitoring large solar systems.[134]

MEASURING BATTERY STATE OF CHARGE

State of charge of a battery is difficult to measure. Voltage is a good indication, but for the best accuracy, the battery must have been at rest for 4 hours. The battery manufacturer can provide the voltage and the specific gravity o the electrolyte corresponding to state of charge (see Table 1).

Table 1 An Example of Voltages for State of Charge[135,136]

% Charged	Cell	12V	24V	32V	48V	Load
100%	2.10	12.60	25.20	33.60	50.40	Open Cell
75%	2.01	12.06	24.12	32.16	48.24	Under Load
50%	1.93	11.58	23.16	30.88	46.32	Under Load
25%	1.84	11.04	22.08	29.44	44.16	Under Load
0%	1.75	10.50	21.00	28.00	42.00	Under Load

The specific gravity of the electrolyte is the most accurate way to measure state of charge and is not dependent on load or charge current. The electrolyte stratifies, so the most accurate measurement is after the gas bubbles liberated during charging have stirred up the acid. The specific gravity is measured using a hydrometer (see Figure 93). The specific gravity must be corrected for temperature (see Table 2).

---

[134] http://www.pvresources.com/en/monitoring.php

[135] http://www.surrette.com/

[136] The load current is not specified. The manufacturer recommends calibrating the table using the specific gravity of the battery acid and doing the test with the normal load. See http://www.surrette.com/ for more details or contact your battery manufacturer.

Figure 93 Hydrometer.

Table 2 Approximate State of Charge Versus Specific Gravity[137]

Approximate State of Charge	40 °F	50 °F	60 °F	70 °F	80 °F	90 °F	100 °F
100%	1.138	1.171	1.204	1.237	1.270	1.303	1.336
75%	1.098	1.131	1.164	1.197	1.230	1.263	1.296
50%	1.063	1.096	1.129	1.162	1.195	1.228	1.261
25%	1.028	1.061	1.094	1.127	1.160	1.193	1.226
0%	0.993	1.026	1.059	1.092	1.125	1.158	1.191
Specific Gravity Readings +/- 0.005							

Charge controllers and other equipment that display battery state of charge have sophisticated algorithms in them that calculate the state of charge based on energy in, energy out, voltage, and temperature. Equipment is available that does nothing but monitor the battery bank (see Figure 94). This type of equipment is handy if you don't have a controller or inverter that displays the battery state of charge or you want to have a backup instrument.

---

[137] http://www.surrette.com/

Figure 94 Battery Monitor.[138]

WEATHER STATIONS

Weather stations are available that measure wind speed, wind direction, and irradiance. Weather stations are normally used by testing facilities that are interested in measuring the effect of wind speed on module temperature. The data is usually recorded using a data logger and stored in a computer.

SUMMARY

If you are working with a system installer, talk to them about instrumentation. The tradeoff is cost versus information. Testing facilities have to have lots of instrumentation to do their job. They use complicated computer programs to analyze the data. A homeowner may only want to know that the system is working OK and what the net energy to the grid is. You can always hire someone to trouble shoot your system if you think there is a problem. The choice is yours.

---

[138] http://www.bogartengineering.com/

14
MAINTENANCE AND REPAIR

In this chapter you will learn how to maintain your system and how to find and fix faults. Keeping your system operating to full capacity is akin to putting money in the bank. Most equipment manuals have a troubleshooting guide in the back that can be most useful in finding a fault. If you run into a problem that you cannot solve, call the equipment manufacturer. If all else fails, you can always hire a professional to fix your system.

COMMON PROBLEMS

Bad Connections

Electrical connections are probably the most common cause of problems, especially in wet or corrosive environments. Temperature cycles and humidity cycles cause slight movements which tend to loosen connections. Battery terminals are a frequent failure point due to battery acid spills and mist (see Figure 95). Connectors are another frequent failure point. Connection problems stem from improper design, improper installation, or improper maintenance. More details about maintaining electrical connections are discussed later on in the chapter.

Figure 95 Corroded Battery Connection.[139]

---

[139] http://strongarmsprays.com/ (StrongArm spray is made to prevent this type of corrosion.)

Soiling and Deposits

Soiling of the modules can cause a significant loss in output (see Figure 96). The average loss due to soiling is 10%, but it can be much more if there are leaves or bird droppings. Only one cell needs to be covered to reduce the module output to almost zero.

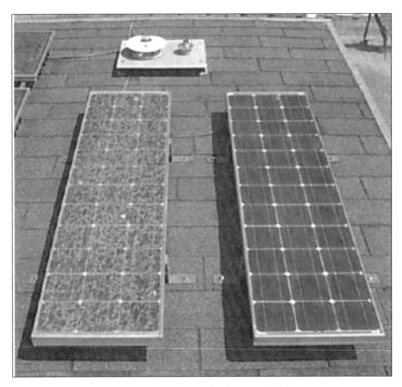

Figure 96 No Rain for a While;
The Module On the Right Has Been Washed Regularly.[140]

Overheating

Overheating of charge controllers and inverters sometimes occurs if it is an extremely hot day with lots of irradiance, if they are placed in the wrong location, or don't have sufficient airflow. The equipment will automatically cut back on output to protect itself.

---

[140] http://www.itfcoatings.com/

Battery Failures

Batteries fail or have reduced capacity for a variety of reasons. (See the section on batteries in this chapter.)

Other Failures

Less common are vandalism, hail damage, modules being blown off the array, structural damage, rodent damage, accidental breakage, and theft. Very occasionally there will be an equipment failure due to a manufacturing defect.

KEEPING IT CLEAN

Rain will usually keep the modules fairly clean. If you live in a dusty area, have lots of pollution or pollen and little rain, you may have to wash the modules with a hose or pressure washer (see Figure 97).

Figure 97 Hosing Off Modules from a Moving Pickup Truck.[141]

---

[141] Courtesy NREL/DOE, Warren Gretz, PVUSA

Any visual soiling will affect the output. Modules on boats sometimes get a lot of bird droppings which must be scrubbed off. Flat-oriented modules may have to be hosed off on a regular basis because rain will cause the dirt to accumulate at the edges.

Some deposits can be hard to remove and require scrubbing. If you decide to scrub your modules, use a mild soap, such as a hand dishwashing soap, and a nonabrasive brush or cloth. Glass cleaner works well. Some modules have a special coating which can be damaged by aggressive chemicals or aggressive scrubbing. If you have questions, consult the manufacturer.

If you have a ground-mounted array, inspecting and washing off the modules can be easy. If your array is roof mounted, you may have to climb up on a ladder. Roof-mounted arrays tend to be more neglected. In some areas you can hire a washing service.

SHADOWS

As mentioned several times, shadows can be real killers, more than you might suspect. Do not allow leaves to remain on your modules. Remove all bird deposits. Trim back any trees that encroach on your array. Do not allow equipment, such as a satellite receiver, to be installed where it will cause a shadow (see Figure 98). If possible, move any equipment that is causing a shadow. If you cannot do anything about a shadow, consider moving the array.

Figure 98 Shadow of Satellite Receiver.[142]

In certain localities you have rights to solar access. If you notice survey stakes on the lot south of your array, go to the building department and ask to see the building permit. Owners are sometimes required to notify neighbors of proposed construction. For the best chance of success, act as soon as possible. You may have to resort to legal action.

ELECTRICAL CONNECTIONS

Electrical connections are the weak link in the wire chain. The author has seen countless failures in various connections, including connectors. Most are caused by improper installation, but a few are caused by a bad connector design or materials. Improper crimping is a frequent cause for connector failure.

---

[142] http://www.mpandc.com/

If you suspect a bad connection, there are several tests to do:

First is a visual inspection. Is the connection or connector discolored? Are wires hanging loose (a sure sign)? Turn the power off and give the wire a gentle tug.[143] Does it come out of the crimped part of a lug? Does it wiggle? Carefully wiggle the insulated part of the wire while the power is on and see if things improve. Use insulated electrician's gloves if there is any chance of being shocked.

Another way to tell is to use your infrared (IR) thermometer while current is flowing. Faulty connections will be warmer than the surrounding area due to the heat generated by increased resistance.

A multimeter is another way to test for a bad connection. For large currents such as battery connections, use it in the voltage setting. Measure the voltage across the connection while current is flowing. Anything more than a fraction of a Volt makes the connection suspect. For small currents, like signal and sensor wires, the resistance setting is more accurate. Anything greater than zero is bad. Test the meter by first putting the probes together. The reading with the probes together is your "zero" reading.

If you find or suspect a bad connection, turn the power off, take it apart, and redo it. If the wire is black, green, or a dull copper color, cut the wire back until you find shinny metallic copper. You may have to replace the entire wire if it got too hot or is corroded. If it is tin-plated wire, cut back until you find shinny metallic tin. Use a new lug or connector. Do a proper job of crimping or tightening (see Chapter 12). Use electrical grease when reinstalling the connections. See more about battery terminals below.

If you find or suspect a bad electronic connector, gently pull the connector apart. Do not pull on the wires. If it does not come apart with a moderate force, check for locking devices. After it is apart, visually inspect the connector. If there are black or green pins or sockets, get a new connector. Use a light viscosity, non-corrosive, oil spray on the pins and sockets, and

---

[143] You want to test the connection without pulling the wire apart, so use a tug that is appropriate for the size of the wire.

Maintenance and Repair

re-assemble the connector. If this does not correct the fault, try wiggling the connector. Sometimes this will make contact. Even if you get it to work, you may want to replace the connector anyway as it is likely to fail again in the future.

Check the tightness of all screws and bolts used to make electrical contact. Use the torque specifications that came with the equipment (see Chapter 12).

Relays have contacts in them that sometimes fail. The resistance and voltage across the contact points when the relay is closed should be close to zero. If you suspect a relay, replace it.

Batteries

Batteries can fail in all sorts of ways: the cases can get cracked, they can sulfate (insulating lead sulfate crystals cover the plates), the plate separators can dry out due to low acid level, the plates can short to each other, or sediment in the bottom can short them out. Most fail by operator error.

You can increase the battery life by following a few simple rules:

- Check the water regularly, at least once a month. If topping up with water is required more than once every two months, the charging voltage is too high.
- Do not overfill the batteries. Fill according to the manufacturer's instructions, usually about a quarter inch below the ring.
- Use only high quality distilled water. The water must not have any metal ions in it.
- Once a month test the specific gravity of a "test cell" after a full charge. Keep records. If the test cell specific gravity starts to fall, consider an equalization charge.
- Once every 6 months check the specific gravity of each cell to see if one cell is starting to fail. Keep records.
- Every 2 to 3 months do an equalization charge according to the battery manufacturer's directions. Do an equalization only on vented batteries.

- Consider using a desulfator to keep the batteries from sulfating.[144]
- Do not allow the batteries to be overcharged or over discharged.
- Keep them at room temperature (50°F to 80°F).
- Clean up any acid spills. Baking soda (bicarbonate of soda) will neutralize any acid. Use 4 ounces per quart of water. Rinse with water and dry.
- Do not allow anything other than distilled water into the cells, including the soda solution.
- Clean any acid mist that condenses on the outside of the batteries.
- Do not short the batteries.
- Do not store the batteries without a trickle charge to keep them charged. Use only a very small amount of current to trickle charge, otherwise you can boil out the water.
- Charge any discharged batteries as soon as possible.
- Do not allow the batteries to freeze.
- Do not allow the batteries to get hot, more than 120°F.
- Use a non-conducting grease, available at battery stores, to protect the terminals and connections.
- Cover the terminals with a plastic or rubber covering to prevent accidental shorting.
- Do not lay tools on top of batteries.
- Do not use a metallic battery tray near the top of the batteries.
- Use a battery box or tray to contain any acid spills.
- On boats, RVs, or other moving locations, use a battery hold-down clamp to keep the battery from shifting. In such installations, use a non-conducting cover to protect the battery.
- Regularly inspect the batteries for any visible damage such as cracks, dents, or deformation.

Battery terminals can build up an insulating layer on them. This is the primary cause of terminal failure and is caused by oxidation of the lead terminals. The terminals may look fine, but will not be conducting. If you find this situation, take the terminals apart and use a special tool or knife to

---

[144] Some people swear by pulse generators, others doubt their value.

scrape the terminals and connectors so that they are shinny. Re-assemble the terminals using the battery grease. Smear grease on the outside of the terminals and wires too. Replace the terminal covers.

Terminals can disappear due to corrosion. This is usually due to acid spills and the effect of current. If the connectors are gone or partly gone, they will need to be replaced. Cut the wire back until it is good. Then clean it with baking soda, rinse with clear water, and dry the wire. Use the battery grease and do a proper installation of the new terminal.

OVERHEATING

If you suspect that the equipment is getting too hot, look at the display, it may have an over-temperature indicator. You can also use the IR thermometer or a thermocouple to measure temperature.

Is the overheating due to an unusually hot day? If the equipment operates normally except for this one hot day, then you may want to just accept that on hot days, you will have reduced output. Check the equipment again in the cool of the next morning. If you are not willing to accept the reduced output on exceptionally hot days, then you can add more shade, increase ventilation, or change equipment.

Some equipment is more temperature tolerant than other equipment. If over-temperature cutback is a real problem for you and you have done all that you can in terms of shade and ventilation, you may have to change to a more temperature tolerant charge controller or inverter. But before you do this, talk to the manufacturer of the existing equipment to see if there is an upgrade or fix.

Modules can get too hot too, normally due to lack of ventilation on the back side. If you suspect hot modules, measure the temperature with the IR thermometer or a thermocouple. Consult the specifications to see what effect this hot temperature has on output. If the figures match, that is your problem. Try to raise the modules up for increased ventilation. Consult with the mounting hardware manufacturer on the best way to do this.

FINDING A FAILED MODULE

Modules do fail occasionally. First do a visual inspection to see if there are any shadows, soiling, loose or broken wires or connectors, or other obvious problems. If possible, look at the IV curve of the array or panel. If there is a funny shape to it, you likely have a failed module (see Figure 99).

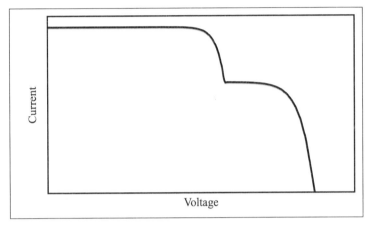

Figure 99 The IV Curve of a 3 Parallel by 3 Series Array With One Failed Module.

If you have multiple source strings, try to isolate the source string with the failed module. Disconnect one source string at a time using the disconnect switch or breaker. Observe what effect removing this one string has on the power. The string with the failed module will have less effect than the others. Pick a day with nearly constant irradiance and do the test in the middle of the day.

To find the failed module in the source string, or array, shadow one module at time and see what effect it has on power. You don't have to shadow the entire module, just a few cells in the string. If the modules have two or more bypass diodes, you will need to use a larger shadow that covers some cells in all strings (see Figure 100). Shadowing the failed module will have less effect on the output than shadowing a good module. Thin-film modules will need a large shadow.

MAINTENANCE AND REPAIR   177

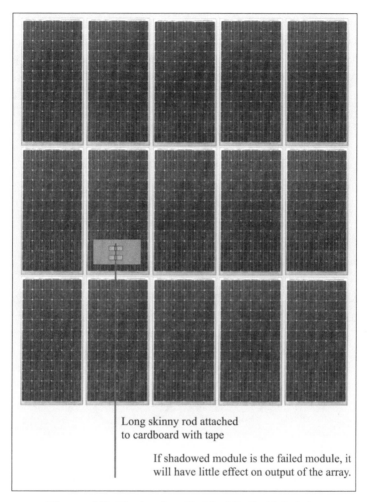

Figure 100 Shadowing To Find Failed Module.

Once you find or suspect a filed module, you will have to test it. If you do not have access to the connectors, you will have to remove the module from the array and test it separately. See below for how to remove modules.

Test the module as described in Chapter 13. Correct the data to standard conditions and compare to the specification. If there is no voltage or no current, you don't have to go any further.

Before you ship the module back to the manufacturer, look it over carefully. Check the wires. Have they come loose from the junction box on the back of the module? Open the junction box (if it will not void the warranty) and look inside. Are there signs of burning? Is the bypass diode fried? Are the leads loose or disconnected? Check the connections with the multimeter. The bypass diode should have high resistance in the non-conducting direction and low resistance in the conducting direction (see Figure 101).[145]

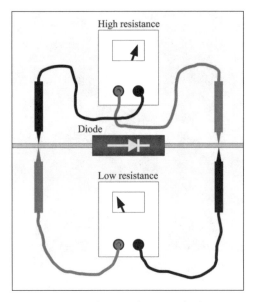

Figure 101 Testing A Diode.

Before you try and fix the module, call the factory and email pictures if possible. The engineers may want to receive the module as found so that they can take corrective actions at the factory.

---

[145] You have to remove the diode before testing. Make sure this will not void the warranty. Consult with the manufacturer on how to remove the diode if you are not sure.

REMOVING AND REPLACING A FAILED MODULE

Removing a failed module can be easy or difficult. If it is in the middle of a large roof mounted array, you will have to walk on the other Modules to get to it. Climbing or stepping on modules is not recommended by module manufacturers but it can be done. Consult with the module manufacturer. Spread your weight out over the module as much as possible. Modules can be slippery. Wear shoes with high traction. Remember to wear your safety harness (see Chapter 15). If possible, walk where the supports are. If it is not possible to walk on the modules, or you do not feel comfortable doing this, you will have to remove modules starting at the edge and work your way to the failed module. Get an assistant to help you.

First disconnect the entire array and all source strings by opening all breakers or disconnect switches. Disconnect the grounded conductor from the grounded module in the source string you are working on. Test to see that you do not have a ground fault that might produce high voltage on the module you are removing. (See Chapters 13 and 15, and Figure 83.)

Do not work on your array if it is windy or extremely hot. Wait until the wind has calmed down and it is cool. The modules can act like sails and get out of control quickly. You can burn yourself on the hot modules.

Most module mounting frames have bolts or screws that are accessible from the front side. Unbolt the failed module at mounting locations and gently lift it up. Unhook the connectors. Do not drop it or allow it to slide on the other modules. You can protect the other modules by taping cardboard to them but make sure that the cardboard will not slip if you step on it. Safety first.

If you have the type of mounting system where the mounting fasteners are not accessible, you will have to remove modules starting at an edge and work your way to the failed module. This is more work, but may be easier and safer than trying to walk on the modules anyway.

The neighbor modules where you have removed the fasteners may be unsupported at that edge. Remount the fastening hardware to hold these modules down while you work on the module you removed. Check to make sure that the type of hardware is OK to use at the edge. If not, get the right edge hardware.

Reverse the procedure to reinstall the module. After everything is back to normal, test the array. If everything is working fine, establish a new benchmark (see Chapter 13).

RECYCLING

Some of the materials used in solar systems are toxic. Lead-acid batteries are toxic. Fortunately the lead-acid batteries can be recycled which reuses most of the lead. Do not leave lead-acid batteries lying around. They can cause injury to any children that come in contact with them and can eventually become an environmental problem. The place where you bought the batteries will recycle them, or search for a site near your location that recycles batteries. The same goes for any of the small batteries that you might use in your flash lights or other equipment.

A good portion of each module can be recycled. The aluminum frame will be bought by a scrap dealer. To remove the frame from the glass, remove any screws at the corners and gently pry the frame away from the glass. Try not to break the glass, which is sensitive to edge stress. The adhesive used to attach the frame to the glass is tenuous, but the frame can be removed. The glass will likely go to a landfill, but check with the module manufacturer first.

Cadmium telluride (CdTe), and copper indium selenide/sulfide (CIGS) modules contain a small amount of material which is toxic. These modules should not go in landfills. Consult with the manufacturer. Most have a buyback or return program for recycling.

The metals used in the mounting frame and the wiring can be recycled. Segregate the metals and take them to the scrap dealer.

15
SAFETY

In this chapter you will learn how to be safe on the job. Accidents can be avoided by following a few simple safety rules. This chapter may be the most important one of the book.

ELECTRICAL SAFETY

Every year several hundred people get electrocuted in the United States. Even so-called safe voltages can hurt you under the wrong circumstances.

Physicals Effects

Electricity can hurt you in three ways. One is electrocution, the second is burns, and the third is involuntary muscle contractions.

Electrocution

Electrocution happens when the electrical current disrupts your nervous system. Your nervous system runs on tiny electrical impulses. A shock can overwhelm that system. The highest danger is when the current passes through the body cavity, shutting down your heart, lungs, or other vital organs. A shock from AC voltage is likely to cause ventricular fibrillation, a kind of quivering of the heart muscle. DC voltage is likely to just stop the heart. Enough of any kind of current will stop the heart. Breathing can be stopped too.

A shock through your brain is another kind of electrocution. The brain shuts down. If you recover, you can have loss of memory or other long-term problems.

Burns

The electrical current can cause first, second, or third-degree burns. Just as a wire will get hot from too much current, your skin and body tissue can be overheated from electrical current. An arc flash, electrical explosion, or a hot wire or tool can also burn you indirectly. The author worked on a 40,000-Watt system that powered a hydrogen generator, at 12 Volts. The

12 Volts was thought to be safe, but when a wrench was accidentally dropped across the two huge bus bars, it literally exploded in a shower of sparks and molten metal.

Involuntary Muscle Contractions

Electrical current causes your muscles to involuntarily contract. If you grab a live wire with your hand, you will not be able to let go. The larger muscles that contract the hand overwhelm the smaller muscles that expand it. The force is sometimes so strong that you can break your own bones. But you are more likely to break a bone from falling after a shock. You will lose control from the involuntary contractions or just be startled enough to lose your balance.

What Voltage is Safe?

Voltage does not hurt you, current does. Your body is like a large resistor. The skin offers more resistance than the tissue. Muscle is more conductive than fat. Wet skin is more conductive than dry skin. Salty skin is more conductive than non-salty skin. The more skin in contact, and the tighter the grip, the more conductive. If you have broken skin, like a cut, you can receive a shock with a 12-Volt battery. Take the two probes of your multimeter, set to resistance, and measure the resistance of your body by holding one probe in one hand and another probe in the opposite hand. If you have dry hands, you might measure 10,000 Ohms. If you hands are sweaty, you might measure 1,000 ohms or less. What effect does squeezing tightly on the probes have?

Treat any voltage as unsafe. The higher the voltage, the more dangerous it is, only because more current will flow.

SAFETY

How Much Current is Harmful

In 1956, C. F. Dalziel did a study on how much current it takes to induce different effects on the human body (see Table 3).[146] He used volunteers up to the point where severe pain was felt, and animal subjects beyond that point. The effect depends on your gender and whether the current is AC or DC. The values in Table 3 are average values and different people respond differently. The values are milliamps (mA), which is 1/1,000th of an Amp. Damage is greater the longer the current persists.

Table 3 Effect on the Body From Different Amounts of Current

Effect	Gender	DC	60 Hz AC
Slight sensation at contact points	Men	1.0 mA	0.4 mA
	Women	0.6 mA	0.3 mA
Threshold of perception	Men	5.2 mA	1.1 mA
	Women	3.5 mA	0.7 mA
Painful, voluntary muscle control	Men	62 mA	9 mA
	Women	41 mA	6 mA
Painful, loss of muscle control	Men	76 mA	16 mA
	Women	51 mA	10.5 mA
Severe pain, difficulty breathing	Men	90 mA	23 mA
	Women	60 mA	15 mA
Possible heart fibrillation after 3 seconds	Men	500 mA	100 mA
	Women	500 mA	100 mA

If you have dry hands, and the total body resistance is 10,000 Ohms, the voltage required to produce 100 mA is 1,000 Volts (0.100 x 10,000 = 1,000). If you are sweaty, you might have a body resistance of only 300 Ohms, in which case only 30 Volts is needed to produce 100 mA (0.100 x 300 = 30). Telephone wires are at 48 Volts. Others have quoted only 50 to 150 mA for possible death.[147] The same source quotes likely death at 1 to

---

[146] Dalziel, C. F., "Effects of Electric Shock on Man," IRE Trans. Medical Electronics, Vol. PGME-5, pp. 44-62, July 1956; also reprinted as USAEC Safety Bulletin 7.

[147] http://www.cdc.gov/niosh

4.3 Amps and probable death at 10 Amps. With higher voltages, more than about 500 Volts, the current actually punctures the skin and the resistance is immediately lowered.

Two Points of Contact

In order to be shocked you must complete a circuit. One part of your body must be in contact with a high voltage while another part is in contact with a lower voltage. The usual situation is one contact on a live wire or electrical component and the other on ground. The ground can be your other hand touching a grounded equipment box or your feet touching the earth. Your shoes provide a certain amount of additional resistance, which is why you should probe with one hand. Shoes, however, are not a safety device, because wet soles can be very conductive. Standing in a pool of water is no protection at all.

A less common connection to two voltages is something called step potential. This happens when there is a high voltage source, such as a downed power line, that is touching the ground. Electricity is flowing from the wire to where ever the power line is grounded. One point is at zero and the other may be at 12,000 Volts. The voltage changes linearly along the surface of the earth. For instance, if the two points are 100 feet away from each other, the voltage changes 120 Volts every foot (12,000 / 100 =120). If you have a stride of 3 feet, every step is a 360-Volt change in potential (3 x 120 = 360). Depending on the resistance of your shoes, you can get shocked just by stepping along the ground. Stay away from downed power lines.

Safe Electrical Work

Here are some rules to live by:
- Assume all voltages are dangerous.
- Never work around electricity alone.
- Learn CPR and get your fellow workers to also learn CPR.
- Do not assume that switches or breakers work. They may be wired wrong or be broken.
- Test for voltage before working.
- Don't assume that your multimeter is working. Test it too.

- Always use lock-outs and tag-outs on switches and make sure <u>all</u> sources to the circuit are locked out (see Figure 102). Test for voltage.

Figure 102 Lock-Out and Tag-Out.[148]

- When working around solar modules, always assume there is voltage if there is any light.
- Use insulated tools such as rubber and plastic handles on screwdrivers.
- Do not assume that any metal object is grounded, there may be a ground fault and the ground wire may be missing or loose. Test for voltage.
- Deplete a circuit that has inductive components or capacitors before working on it. Test for voltage.

---

[148] http://kvazar.hu/

- Inform everyone on the job when re-energizing circuits. Only authorized persons should re-energize circuits.
- When working on high voltage, use electrician's gloves.
- Wear a non-conducting hard hat to keep from accidentally getting shocked through your skull.
- Use non-conducting ladders around electricity, not aluminum. Wet wood is conductive.
- Do not work near overhead lines, because they are not insulated.
- Minimum distance from a 50,000-Volt overhead line is 10 feet, more for higher voltage.
- Do not dig near buried electrical lines.
- Do not use a fish tape on a live system, it can cut through the wire insulation.
- Do not use ungrounded, unsafe, or broken equipment or tools.
- Do not use frayed extension cords or tools with frayed wires.
- Do not incorrectly wire electrical systems. Follow the color-coding: green or bare for equipment ground, white for grounded conductor, black or other color for non-grounded conductor.
- Do not assume that the white wire is grounded and the black or red wire is hot, they could be incorrectly wired. Test for voltage.
- Never work on a live circuit. Test for voltage.
- Never work when tired, angry, in a hurry, emotionally upset, or intoxicated.
- Never work when you are taking medication that causes drowsiness or using drugs.
- Work in a clean, uncluttered environment.
- Work slowly and deliberately.
- When working over 6 feet off the ground, use a fall restraint.
- Have a class C fire extinguisher on site.
- If you find a co-worker being electrocuted, to prevent becoming the second victim, turn off the power first or move him away from the source with an insulating tool like a piece of wood. Check for consciousness, breathing, heart function, and administer CPR if necessary. Have someone call for emergency medical services (911).

SAFETY

- Do not overload circuits.
- Do not work in wet conditions.
- Wear proper clothing and protective gear such as eye protection.
- Do not use sprays near live electrical components, they can conduct electricity or catch fire.
- Learn to recognize hazards and take corrective action.
- Use ground fault interrupters (GFIs). Portable GFIs are available for use on extension cords.
- Switches and breakers must be clearly labeled.
- Do not use the wrong extension cord or one that has the wrong ends.
- Do not use extension cords where they might be damaged or kinked.
- Use the correct wire size and type for the job.
- Maintain equipment that restricts access, replace covers, and guard against unauthorized contact.
- Maintain a proper grounding system.
- Remove any jewelry before working on electrical equipment.
- Make sure breakers and fuses are sized correctly.
- Use a fuse or breaker on all sources of power.[149] If using a fuse, there must be a means to de-energize the fuse to service it. Test the fuse for voltage before servicing.
- Do not allow casual visitors or children into a work site.

---

[149] The NEC allows exceptions.

A good manual for electrical safety, "Electrical Safety, Safety and Health for Electrical Trades, Student Manual," is supplied free by the:

> Department of Health and Human Services
> Centers for Disease Control and Prevention
> National Institute for Occupational Safety and health

Order at:

> NIOSH-Publications Dissemination
> 4676 Columbia Parkway
> Cincinnati, Ohio 45266-1998
>
> 1-800-356-4674
> **www.cdc.gov/niosh**

BATTERY SAFETY

Besides the electrical safety issues discussed above, batteries offer a number of additional safety hazards. Battery acid is very corrosive to materials and also to skin and eyes. Batteries store enormous amount of energy that can cause fires, burns, and melt tools. Under the right circumstances, the batteries can explode from overheating. During charging they generate hydrogen and oxygen gases that are explosive in combination and the hydrogen alone is explosive in air.

Electrical Safety for Batteries

Observe all the electrical safety rules discussed above. Open all the switches and breakers to de-energize the system. Use lock-out and tag-out procedures. *In addition, follow these rules:*

- Do not place tools on top of batteries or place tools where they can drop on batteries.
- Keep metallic frames and other metallic objects at least 6 inches away from the battery terminals.
- Do not connect or disconnect the battery terminals under load, make sure the disconnect switch or breaker is open. Test the voltage downstream of the disconnect.

- Verify the polarity when hooking up batteries.
- Never short a battery, it can explode.
- Connect the battery side of cables first, and then connect the controller, inverter, or load center side of the cable.
- Use conduit as required by the NEC.
- Use disconnect switches or breakers that are switch rated.
- Use equipment DC rated for the voltage and current.
- Never work alone. A helper should be available for immediate assistance if needed.

General Battery Safety

Batteries here refer to lead-acid batteries that contain sulfuric acid. It is one of the strongest acids in general use.

Battery handling rules:

- Wear old clothes that you don't mind getting holes in.
- Wear a protective rubber apron, protective rubber gloves, and a face shield.
- Keep fresh water nearby to flush acid spills. A fresh-water hose is required to flush your eyes in case they get splashed with acid.
- Have on hand baking soda (bicarbonate of soda) solution for neutralizing acid spills. Use 4 ounces of baking soda to 2 quarts of water. After neutralizing, flush with water and dry.
- In the event of exposure to acid, wash the affected area with soap and water. If your eyes get splashed, wash with running water for 15 minutes and seek immediate medical attention.
- Use battery trays to catch acid spills.
- Use proper lifting techniques. Batteries are heavy; some require special lifting equipment. Do not lift by the terminals.
- Keep battery rooms locked to keep casual visitors and children away from the batteries.
- The battery room should be well ventilated with the vents at the top.
- Never smoke in the battery room or allow any open flames.

- Use protective grease on the terminals to keep acid from corroding them.
- Cover the terminals to prevent accidental contact with metallic objects (see Figure 103).

Figure 103 Battery Terminal Covers.

- Use batteries that are identical in size, type, manufacturer, and date. Check the date code.
- Never add used batteries or batteries of unknown condition.
- Do not overcharge batteries.
- Keep the battery room clean and uncluttered.
- Never work on batteries when you are tired, emotionally upset, intoxicated, on medication that causes drowsiness, or using drugs.
- Post the battery room with caution signs indicating high voltage and the presence of acid.
- Recycle used batteries and never leave them lying around where children or others can get to them.

WORKING ON ROOFS OR AT HEIGHTS

Any time you are working above 6 feet, you should be wearing a fall arrest system. This is usually a harness attached to a shock-absorbing tether (see Figure 104).

Figure 104 Fall Arrest System.[150]

---

[150] http://www.meico.cz/

The tether should be attached to a secure object such as an anchor at the peak of the roof (see Figure 105).

Figure 105 Fall Arrest Anchor.[151]

On tall roofs, the tether should prevent you from reaching the edge of the roof. Falls are the leading cause of fatalities in the US construction industry.[152]

---

[151] http://www.rd.com/how-to-properly-use-a-roof-safety-harness/article18126.html
[152] http://www.osha.gov/doc/outreachtraining/htmlfiles/subpartm.html

Rules for working safely on roofs or at heights:

- Use a proper fall arrest system.
- Make sure the fall arrest system is in good condition and follow the manufacturer's instructions.
- Anchor the fall arrest system securely to roof rafters per the manufacturer's instructions.
- Wear shoes that have good traction.
- Sweep the roof free of debris, leaves, loose shingle granules, and construction materials.
- Keep the roof clear of tools and other obstructions that might cause you to trip.
- Never work at heights when you are tired, emotionally upset, intoxicated, on medication that causes drowsiness, or using drugs.
- Use a proper ladder of correct size and secure it to the roof edge using rope or wire tied to heavy-duty hardware screwed or nailed to the rafter ends or fascia.
- Extend the ladder 3 feet above the roof to provide a handhold when ascending or descending.
- Use both hands on the ladder when using it.
- Use a helper to lift tools and materials onto the roof.
- Set the ladder on firm, level ground and at the proper angle.
- Do not allow anyone to stand below tools and materials that are being lifted onto a roof.
- Do not work on wet or hot roofs. Pick a cool, dry day or work in the morning before it gets hot.
- Do not work on roofs on windy days, wind can make modules unmanageable.
- Keep others away from below where you are working by roping off the area, using safety tape, or using safety monitors on the ground.
- Before tossing anything off the roof, assure that there are no persons below.
- Keep extension cords away from where you are walking.
- Tie off tools and materials to keep them from falling.
- Stay off loose shingles or shingles that can come loose or are slippery.

- Do not overload the roof with materials; spread out the materials to distribute the load.
- If you intend to spend any time on the roof, install roof jacks (also called roof brackets) (see Figure 106).

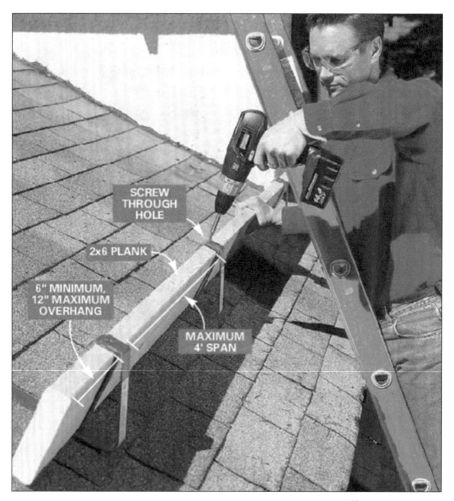

Figure 106 Roof Jacks with Plank Installed.[153]

---

[153] http://www.rd.com/how-to-properly-use-a-roof-safety-harness/article18126.html

- Nail roof jacks to rafters following the manufacturer's instructions.
- Steep roofs require special precautions such as tiers of roof jacks or ladders secured to the roof surface.
- Keep off skylights or other structures that may not support your weight.
- Use a guardrail around any roof openings.
- If you use a scaffold, assure it is assembled correctly, on solid ground, and secure.
- If you are using machinery to lift materials to the roof, such as a forklift, make sure it is operated in a safe manner and operated on firm, level ground. The operator of machinery should be trained in its proper use.
- Wear hard hats, especially if you are the helper on the ground.
- Wear a protective sunscreen and protective clothing to keep from getting sunburned.
- Keep hydrated by drinking lots of water.
- Take breaks. If you feel like you are starting to suffer from heatstroke, move off the roof and into the shade and seek medical attention.

Working Safely with Hand Tools

Any hand tool can cause an injury, especially powered tools. Most people who have worked with a hammer for any length of time have smashed a thumb. You can lose a finger with a power saw, stab yourself with a screwdriver, injure an eye with flying debris, or cut yourself with a mat knife. The list goes on. But these injuries don't have to happen if you work carefully and follow a few simple safety rules.

Rules for working with hand tools:
- Never work with hand tools when you are tired, emotionally upset, intoxicated, on medication that causes drowsiness, or using drugs.
- Read and follow the instructions that came with the tool.
- Wear safety equipment, including eye protection, ear protection, and any other personal protection equipment that is appropriate.

- Wear a mask when working around sawdust or other airborne particulates.
- Do not wear loose-fitting clothing and remove jewelry.
- Make sure that safety guards are installed before using a tool.
- Do not defeat or remove safety guards.
- Work carefully and do not rush the job.
- Keep your work area clean and clear of clutter.
- Treat power nailers like a lethal weapons; never point them at anyone or treat them like toys.
- Avoid horseplay, practical jokes, or other inappropriate behavior.
- Inform others on the job that you are using power nailers or powder actuated tools.
- Keep hands and other body parts away from the business end of power nailers.
- Keep tools sharp, clean, and in good condition.
- Use the right tool for the job and do not use tools in a manner for which they were not designed.
- Secure the work with a clamp, vice, or other means.
- Cut away from yourself.
- Do not hammer on a hardened steel surface, for instance on another hammer, as chips can fly off and cause injury.
- Do not carry sharp tools in you pocket.

Working on projects does not have to lead to injury if you work safely. Safety is an attitude. Injuries occur in moments of inattentiveness or when trying to take shortcuts.

GLOSSARY

Absorbent Glass Mat (AGM) – A lead-acid Battery with fiberglass mats between the plates to contain the acid

Absorption Charge – The second-stage charge in a Three-Stage Charge Controller

AC – Alternating Current

AGM – Absorbent Glass Mat battery

Ah – Amp-Hours

Alternating Current (AC) – Current in an electrical Circuit that is rapidly flowing in alternating directions

Alternative Energy – Any source of energy except fossil fuels and nuclear energy

Amorphous Silicon (a-Si) – A Thin-Film Module made with the amorphous (non-crystalline) form of silicon

Amp-Hours (Ah) – A unit of measure for the capacity of a Battery or Battery Bank

Annealed – A heat treatment process that changes the properties of a material

Apparent Power – The current times the voltage of an AC circuit that includes inductive loads, measured in Volt-Amps, see Real Power and Power Factor

Aquifer – An underground water source

Array – In this book, a Solar Array

a-Si – Amorphous Silicon Module

Back Contact Solar Cells – Solar Cells that have both their negative and positive contacts on the back side

Balance-of-system (BOS) – The parts of a Solar Electric System besides the Solar Modules

Benchmark – A test done when a system is new so that performance can be verified and later changes can be determined

Battery – Two or more electrochemical cells that store and release electricity through chemical energy, or by common usage one electrochemical cell such as an "AA battery"

Battery Backup – A system where a Battery Bank is used to supply power when the primary source is not available

Battery Bank – Two or more Batteries joined together to add capacity and/or voltage

Battery Desulfator – A device designed to convert lead sulfate crystals into the amorphous form

Blocking Diode – A diode used to prevent reverse current flow through a Source Circuit or single Module

Bonds – Sold by state governments and utilities to raise money for alternative energy projects

Brushless Motor – A motor that does not use brushes to transfer current to the armature

Bulk Charge – The first stage charge in a Three-Stage Charge Controller

Bypass Diode – A diode added in parallel with a number of Solar Cells to allow current to flow around the cells when there is a fault or shadow

C rate – The rate of charging or discharging of a Battery measured as a fraction of the capacity of the Battery in Amps such as C/5

Cadmium Telluride (CdTe) Module – A Thin-Film Solar Module made using a compound of cadmium and tellurium

Casing – A pipe placed in a water well to prevent it from caving in and to seal the Aquifer from surface contamination

Centrifugal Pump – A water pump using one or more rotating impellers to pump the water

Certification – Verification by an independent organization that a piece of equipment meets certain safety, quality, or performance requirements

CdTe   Cadmium Telluride Module

Charge Controller – A device that controls the amount of current sent to a Battery Bank

CIGS – Copper Indium Selenide

Circuit – An electrical connection that allows current to flow in a loop

Circuit Breaker – An over-current protection device that disconnects a load or device when the current is excessive

Combiner Box – A box that combines the output of a number of Solar Modules

Concentrating Module – A Solar Module that uses optical elements, such as mirrors or lenses, to concentrate the sunlight onto the Solar Cells

Copper Indium Selenide (CIGS) – A Thin-Film Solar Module made from a compound of copper, indium, gallium, and selenium

Cost of Electricity – The cost of electrical energy in units of cents per kilowatt-hour
Credits – Funds that are deducted from the cost of a system such as tax credits
Crystalline Silicon – Silicon with a regular ordered molecular structure
Data Link – Hardware and software for transferring data in and out of a piece of equipment
Data Logger – A system or device used to measure and store performance data
Dead Load – The load on a roof that does not move, such as the rafters and sheathing
Deep Discharge Battery – A lead-acid Battery designed for long and deep discharge cycles
Depth of Discharge (DOD) – The percentage of energy taken out of the battery, see State of Charge
Direct Current (DC) – Current in an electrical Circuit that is always flowing in the same direction
Discharge Curve – The discharge characteristic of a pump expressed as a curve
DC – Direct Current
Direct-Connect Solar Pumping – A Solar Powered Water System where the Solar Array is directly connected to the water pump without a Battery Bank
Disconnect – A switch for disconnecting a load or power source
Diversion Charge Controller – A charge controller that diverts excess power to control Battery voltage
DOD – Depth of Discharge
DSIRE – Database of State Incentives for Renewables and Efficiency, an up-to-date database of state, federal, and local incentives for energy efficiency and alternative energy, **http://www.dsireusa.org/**
Dynamic Head – The pressure in a water system due to the friction of the water moving in the pipes and fittings
Efficiency – A measure of how much power is supplied for a given amount of input power, the ratio of output power divided by input power
Electrolysis of Water – DC electricity is used to separate the hydrogen and oxygen contained in the water molecule

Energy Audit – An examination of the energy efficiency of a house or commercial building and a recommendation of improvements

Equalization Charge – An intentional overcharging of a Battery to remove the contaminates from the plates and equalize the State of Charge of all the cells

Equipment Ground – Conductors that connect the electrically conductive parts of equipment to a ground terminal

Feed-in Tariff – Purchase rate paid by utilities for alternative energy that is above the market rate

Fill Factor (Ff) – A measure of the quality of the IV Curve, (Vmp X Imp) / (Voc X Isc)

Fixed Tilt – A Module, Panel or Array that is fixed so that it does not move; not mounted on a Tracker

Flat-Plate – A Solar Module that is not a Concentrating Module

Flexible Impeller Pump – A pump using distortion of a rotating flexible impeller to pump water

Float Charge – The third stage charge in a Three-Stage Charge Controller

Flooded Cell – A Battery that uses liquid electrolyte

Foot Valve – A check valve located in the bottom of pipe to keep the pipe full of water

Fuel Cell – A device for combining hydrogen and oxygen from air to produce electricity

Gallium-Arsenide (GaAs) – A Solar Cell made from the crystalline form of gallium and arsenic

Gel Battery – A lead-acid Battery in which the acid is in the form of a gel, also called a gel cell

Global Irradiance – Irradiance coming from all directions of the hemisphere perpendicular to the plane of measurement

Grants – Gifts of funds for worthwhile projects, usually to small businesses, educational entities, and government entities, see DSIRE

Grid – The system that distributes power from the utility companies to the users of the power

Grid-Connected Inverter – An Inverter that is designed to be used in conjunction with a grid

Grid-Connected System – A Solar Electric System connected to a utility company power Grid; power flows to and from the Grid

Grid-Tied – Grid-Connected System

Ground Fault Circuit Interrupter (GFCI, GFPD, GFI) – A device that disconnects a circuit when current flows through an Equipment Ground conductor

Ground Fault Interrupter – See Ground Fault Circuit Interrupter

Ground Fault Protection Device – See Ground Fault Circuit Interrupter

Head – A term used for pressure in a water pumping system, measured in feet or meters

Helical Rotor Pumps – A Positive Displacement Pump using a helical rotor in a helical housing to pump the water, also called a progressing cavity pump

High Temperature Derating – A protective system used in Inverters and Charge Controllers to cut back the output when the equipment is overheated

High Voltage Disconnect (HVD) – A device to disconnect a Solar Array from a Charge Controller or Inverter when the Array Voltage is too high

Horsepower – A measure of power, equivalent to raising a 33,000-pound weight one foot in one minute

Hybrid Solar System – A system that includes a Solar Electric source and another source such as a wind turbine, hydro turbine, or engine-driven generator

Hybrid Inverter – An Inverter that operates as a Grid-Connected Inverter when the Grid is present, but disconnects from the Grid and operates as a Stand-Alone Inverter when the Grid is offline

Hydrometer – A device used to measure the specific gravity of the electrolyte in a lead-acid Battery

Imp – Peak Power Current

Infrared (IR) Thermometer – A device for measuring temperature at a distance using infrared radiation

Initial Efficiency – The Efficiency of an Amorphous Silicon Module when first made, see Stabilized Efficiency

Irradiance – The rate of Solar Radiation in Watts per square meter

Intangible Credits – Credits for using renewable energy that are not directly related to the financial aspects, such as environmental benefits

Inverter – A device for converting DC current to AC current

Isc – Short Circuit Current

Islanding – A condition where a Grid-Connected Inverter continues to run even though the power from the utility company is absent

IV Curve – A plot of current versus voltage for a Solar Cell, Module, Panel, or Array

Jack Pump – A Piston Pump where the motor and gearbox are at the surface of a well and the pump is at the bottom and connected to the motor by a reciprocating rod

Jet-action Pump – A Surface Pump where part of the water is pumped to below the water level to force water to the pump by the action of a jet

Junction – The boundary between the P-type and N-type parts of a Solar Cell

Kilowatt (KW) – One thousand Watts

KW – A Kilowatt

LED – Light-emitting diode

LCD – Liquid crystal display

Life-cycle Cost – The total cost of a system during its System Lifetime

Live Loads – The loads on a roof that change, such as wind loads or the weight of people walking on the roof

Load Center – A box that houses Circuit Breakers or fuses that supply power to loads in a residence or business

Loans – Low-interest loans are available for energy efficiency improvements and for alternative energy projects, see DSIRE

Low Voltage Cutoff (LVC) – See Low Voltage Disconnect

Low Voltage Disconnect (LVD) – A device that disconnects a Battery from a load when the Battery Voltage is low, also called Low Voltage Cutoff

LVC – Low Voltage Cutoff

LVD – Low Voltage Disconnect

mA – Milliamp

Maximum Peak Power Tracking (MPPT) – A system in which an Inverter or Charge Controller keeps the Solar Electric System operating at its Maximum Power

Megawatt – A million Watts, a thousand Kilowatts

Milliamp (mA) – 1/1,000th of an Amp

Millivolt (mV) – 1/1,000th of a Volt

Modified Square Wave – An AC wave form consisting of zero voltage and single value positive and negative voltages

GLOSSARY 203

Module – A basic building block; in this book it means a Solar Electric Module

MPPT – Maximum Peak Power Tracking

Multiple Junction – A Solar Cell with more than one Junction, also called multi-junction

MW – A Megawatt

mV – Millivolt

N-type – The negative part of a Solar Cell

Off-Grid – Not connected to the Grid

Open Circuit Voltage (Voc) – The voltage of a Solar Cell, Module, Panel, or Array when the wires are not connected to a load

Over Current Protection Device – A fuse or Circuit Breaker designed to disconnect a circuit in the case of excessive current

Ozone Generator – A device that generates ozone for use in sanitizing water

Panel – In this book, a Solar Panel

Parallel Connected – Equipment wired to add current with all the negative poles connected together and all the positive poles connected together

Payback Period – The time required for the Solar Electric System to operate in the black

Peak Power (Pmax) – The maximum power produced by a Solar Cell, Module, Panel, or Array

Peak Power Voltage (Vmp) – Voltage at Peak Power

Peak Power Current (Imp) – Current at Peak Power

Photovoltaic – Electrical energy from sunlight using solar cells

Piston Pump – A Positive Displacement Pump that uses pistons to pump the water

Pmax – Peak Power

Polycrystalline Modules – Solar Modules in which the Solar Cells have multiple crystals

Positive Displacement Pump – A pump that traps a constant volume of water and pumps that volume for each revolution of the shaft

Power Factor – The ratio of Real Power to Apparent Power

Pressure Tank – A tank that uses air pressure to supply water when the pump is not running; used to reduce the amount of start and stop cycles of the pump

Production Incentives – Payments for alternative energy production, see DSIRE

P-type – The positive part of a Solar Cell

Pulse Width Modulation (PWM) – A type of current control using voltage pulses of varying widths

PV – Abbreviation for Photovoltaic

PWM – Pulse Width Modulation

Pynanometer – An instrument for measuring Irradiance

Real Power – Power that does useful work, measured in Watts

Rebates – A financial incentive in the form of funds to pay for part of energy efficiency improvements or a Solar Electric System

Recurring Costs – Costs that are paid during the System Lifetime and not part of the initial costs

Renewable Energy – Energy from sources that are naturally replenished

Reverse Osmosis – Desalination of water using high pressure to force the water through an osmotic membrane

Resistance Temperature Detector, (RTD), A device that measures temperature by using a metallic resistive element

Series Connected – Equipment wired to add voltage with the positive pole of one wired to the negative pole of the next

Short Circuit Current (Isc) – The current from a Module, Panel, or Array when the wires are shorted together; current when the voltage between the plus and minus terminals is zero

Sine Wave Voltage – An AC circuit where the voltage follows a sine wave

Single Crystal Modules – Solar Modules in which the Solar Cells are made from single silicon crystal

SOC – State of Charge

Solar Array – A collection of one or more Solar Panels

Solar Cell – A semiconductor device that converts sunlight to electricity

Solar Electric Energy – Electricity generated from sunlight using Solar Cells

Solar Electric Module – A sealed unit with one or more Solar Cells, used to generate electricity from sunlight

Solar Energy – Energy from sunlight; in this book it means Solar Electric Energy

Solar Panel – A collection of one or more Solar Modules

GLOSSARY

Solar Powered Water System – A system where Solar Electric Energy is used to pump water

Solar Radiation – Radiation coming from the sun, sunlight

Source Circuit – A Series Connected group of Solar Modules

Square Wave – An AC wave form consisting of alternating single value positive and negative voltages

Stabilized Efficiency – The Efficiency of an Amorphous Silicon Cell after 1000 hours in sunlight, lower than Initial Efficiency

Stacking – Two Inverters are connected in Series to increase output voltage

Stand-Alone Battery System – A solar system that is not Grid-Connected and uses a battery bank to store the solar energy

Stand-Alone Inverter – An Inverter designed to be used without the presence of the Grid

Standard Test Conditions (STC) – Test conditions corresponding to 1000 Watts per square meter Irradiance and 25°C Junction temperature

Starting Battery – A lead-acid Battery designed to start engines

Starting Current – Surge Current

State of Charge (SOC) – The percentage of charge in a Battery

Static Head – The pressure in a column of water when the water is not moving

STC – Standard Test Conditions

Submersible Pump – A pump designed to be used below the surface level of the water

Suction Head – The suction necessary to keep a static column of water in the suction side of a pump

Surface Pump – A water pump designed to be used without being submersed in the water.

Surge Current – A larger than normal current present when certain types of motors or equipment are started, also called Starting Current

System Lifetime – The operation life of a system

Tandem Cell – A Solar Cell with two Junctions

Tax Breaks – Reductions in taxes for energy efficiency or renewable energy equipment

Tempered Glass – Glass that has gone through a heat tempering process to improve its strength and safety

Terrestrial – On the earth

Thermocouple – A device used to measure temperature using the voltage difference between two junctions of different metals at different temperatures

Three-Stage Charger Controller – A Battery Charge Controller designed to charge at three different rates depending on the State of Charge of the Battery

Thermistor – A device that measures temperature by using a semiconductor resistive element

Thin-Film Module – A Solar Module made with a thin film of active material; see Amorphous Silicon, Cadmium Telluride, and Copper Indium Selenide

Tracker – A machine that moves to point a Solar Panel towards the sun

Transformer – A device for increasing or decreasing the voltage of an AC system

Total Dynamic Head – The sum of Static Head, Dynamic Head and Suction Head

Utility Grid – See Grid

UV Water Treatment – The process of using ultraviolet (UV) radiation to sanitize water

VAC – Average Voltage of an Alternating Current Circuit

Valve Regulated (VRLA) – Lead-acid Batteries that have a pressure valve to prevent escape of hydrogen and oxygen gas during normal charging

VDC – Voltage of a Direct Current Circuit

Vmp – Peak Power Voltage

Voc – Open Circuit Voltage

Volt (V) – A measure of electrical potential

Voltmeter – An instrument for measuring voltage

W – A Watt

Watt (W) – A measure of electrical power

Watt-hour (Wh) – A measure of electrical energy, one Watt for one hour

Weep Hole – A small hole used to drain water from a pipe but not large enough to affect normal water flow

Weighted Efficiency – The efficiency of a device that is averaged over the normal operating regime of the device

Wh – A Watt-hour

INDEX

applications, 19-30
    grid-connected solar systems, 1, 19-21, 26, 43, 44, 45, 48, 53, 54, 108, 118, 124, 127, 128
    grid-connected with battery backup, 26, 48, 64
    hybrid solar systems, 24, 25, 26, 27, 48, 57, 72, 73, 103, 163
    hydrogen fuel generation, fuel cells, 30, 73, 182
    other applications, 29
    space and satellites, 1, 38
    stand-alone (off-grid) solar systems, 18, 22, 43, 46, 47, 48, 52, 108
    water pumping, 27, 28, 29, 79-98, 103, 106, 146
batteries, 55-68
    additives, 67
    aging, 67
    basics, definitions, 55
    capacity of a battery, 55
    charging batteries, 14, 22, 23, 26, 46, 48, 5, 55, 56, 59, 61, 64, 65, 66, 69, 70, 71, 73, 74, 103, 125, 162, 163, 164, 165, 166, 172, 173
        3-stage charging, 65, 66, 70, 71
        blocking diode, 21, 69, 70
        C rate, 55
        charging voltage, 74, 78, 173
        depth of discharge (DOD), 56, 68
        direct connection to array, 21, 69, 70
        equalization charge, 59, 66, 67, 74, 173
        pulse width modulation, 72
        simple charger, 65, 70
        specific gravity of electrolyte, 164, 164, 173
        state of charge (SOC) 65, 66, 67, 162, 164, 165
        temperature compensation, 66, 74
    choosing batteries, 68
    desulfator, 67, 174
    expense, 22, 26, 64, 71
    ground fault circuit interrupters (GFCI), 51, 77, 105, 187
    installing, 64, 138, 173, 188
    lead-acid batteries types, 56, 57-62
        absorbent glass mat (AGM), 60, 66, 74
        deep discharge batteries, 62
        flooded cell, 57, 58, 59, 60, 61, 62, 64, 65, 66, 74

        gel battery, gel cell, 61, 66, 74
        hybrid batteries, 62
        starting battery, 22, 57, 62
        valve regulated, 59, 60, 61, 66, 68, 74
    lifetimes and extending lifetime, 57, 62, 66, 67, 125, 137
    low voltage disconnect, 47, 65, 74, 78
    maintenance, 57, 58, 59, 64, 65, 164, 165, 167, 173, 174, 175
    other types of rechargeable batteries, 56, 57
    performance, 22, 62
    safety, 22, 56, 59, 60, 61, 64, 68, 173, 180, 188-190
    temperature, effect of, 66, 67, 74, 125, 138, 164, 174
    testing batteries, 164, 165, 173
    water recombining caps, 59
    wiring, parallel, series, 63, 64, 65, 68
California Energy Commission (CEC), 45, 49, 118
charge controllers, 22, 56, 65, 66, 69-78, 160, 164
    certifications, 78
    charging voltage, 74
    choosing a charge controller, 70, 73, 78, 103
    diversion charge controller, 72
    efficiency and performance, 70, 72, 76, 78
    features, 70-78
    ground fault circuit interrupters (GFCI), 77
    high voltage disconnect (HVD), 75
    installation, 77, 138, 139, 141-148
    low voltage disconnect (LVD), 74
    maintenance, 167
    maximum power point tracking, 45, 70
    parallel connection for more current, 75
    primary function, 69
    pulse width modulation, 72
    temperature of battery compensation, 66, 74
    temperature derating, 74
    types of charge controllers, 70
        built into inverter, 52
        diversion charge controllers, 72
        multiple-source controllers, 73
        simple charge controller, 70
        three-stage controllers, 65, 70
cost of electricity, 127
    cost of electricity from a generator, 128

INDEX

    cost of solar-electric generated electricity, 127
    credits, 126
    discounted cash flow, 123
    eliminating or reducing electricity bills, 1, 19, 20, 128, 129
    intangible credits, 129
    life-cycle cost, 125-127
    lifetime of equipment, 125
    payback period, 109, 128
    production incentives, 112
    system cost, 124
    types of cost calculations, 123
designing a solar system, 99-106
    balance of system (BOS), 34, 106, 131-148
    battery sizing, 24, 26, 55, 56, 62, 64, 68, 103, 125
    choosing equipment, 103
    documentation, 106
    electrical loads, 24, 26, 43, 44, 47, 50, 72, 101, 129
    energy audit, 99
    foundations, 136
    losses and low output, 7-13, 29, 34, 44, 67, 77, 94, 99, 100, 102, 103, 104, 138, 154, 167-173, 175
    shadows, 8, 9, 16, 36, 104, 147, 168, 169, 170, 176
    site survey, 104, 131
    structural considerations, 16, 34, 38, 97, 104, 105, 131-137, 169
    wiring, breakers, disconnects, 20, 21, 23, 29, 44, 47, 50, 51, 52, 63, 64, 65, 68, 69, 74, 75, 102, 106, 125, 139-146, 171-173
financial incentives, 107-114
    bonds, 114
    DSIRE, 107
    feed-in tariff, 112
    federal incentives, 107
    getting your state to adopt incentives, 114
    grants, 112, 113
    loans, 109, 111
    rebates, 108, 109, 110, 127
    state incentives, 108
    tax breaks, credits, 112, 126
installing systems, 131-148
    animals and insects, protecting against, 145-147
    attachment, 131-137
    batteries, installing, 138

conduit, 145
connectors, installing, 139-144
corrosion, protecting against, 141-144
foundations, 136
ground-mounted, 136
grounding, 143
handling modules, 18, 139, 141
holes, drilling in module frames, 141
loading, structural, 131-137
locating, 77, 137
mounting hardware, 77, 130, 131
preventing abrasion, 139
quality workmanship, 148
roof, installing on, re-roofing, 131
security, 148
snow loads, 137
solar access rights, 171
thermal considerations, 138
UV radiation damage, protecting against, 144
wiring, 139-144
instrumentation and testing, 147-164
    ammeter, clamp-on 153
    basic instrumentation, 150
    batteries, measuring state of charge, 162-164
    data exchange, 50, 52, 77, 162, 163, 164
    data logging, 50, 162, 163, 164, 166
    digital multimeter, 149-152
    IV curve, measuring, 159-161
    IV curve tracer, 161
    pynanometer, 154-156
    standard conditions (STC), correcting to, 161
    sunlight, measuring, 154-156
    temperature, measuring, 156
        infrared (IR) thermometer, 157
        other means to measure temperature, 158
        thermocouples, 156
    weather stations, 166
inverters, 43-54
    battery charger built in, 52
    certifications, 52
    choosing an inverter, 54, 103

efficiency and performance, 49
features, 50-52
ground fault circuit interrupters, 51
high temperature derating, 50
hybrid inverters, 48
islanding, 44
low-voltage disconnect (LVD), 47
maximum power point tracking, 45
power factor and VA capacity, 51
stacking, 50
surge current, 48
types of inverters 43
    grid-connected (grid-tied) inverters, 19, 43
    stand-alone inverters, 22, 46
wave forms, sine wave, square wave, 44, 45, 47

maintenance, 167-180
    batteries, 64, 65, 164, 173-175
    common problems, 167
    connectors and connections, 171-175, 178
    module failure, 176-180
        examining a failed module, 177, 178
        finding a failed module, 176
        removing a failed module, 179
    recycling, 180
    shadows, keeping them off of arrays, 170, 171
    soiling, 169, 170
    thermal issues, 175

National Electric Code (NEC), 51, 52, 105, 143, 189

safety, 181-196
    battery safety, 188
        safety rules, 188, 189
    electricity, 181-188
        burns, 181
        electrocution, 181-184
        involuntary muscle contractions, 182
        more information, 188
        safety rules, 184
        what is a safe voltage, 182
    ground faults, 51, 105, 179, 187
    ground fault circuit interrupters (GFCI, GFPD, GFI), 50, 51, 52, 77, 78, 105, 187

hand and power tools, 195
    safety rules, 195
roofs and heights, 191-195
    fall restraints, 191, 192
    safety rules, 193
solar cells, 31
    junction, 31, 38
    how solar cells work, 31
    thin-film solar cells, 35
solar modules, 31-41
    bypass diodes, 16
    choosing modules, 105
    definition of a solar electric module, 4, 7
    gallium-arsenide (GaAs) cells, 38
    how solar modules work, 8, 31
    indoor applications, 38
    integrated into roof sheathing, 36, 38, 135
    IV curve and points of IV curve, 14, 151, 159-162
    labels, 14, 15
    output and performance, 31-41
        clouds, climate, and weather, 10, 11, 24, 37, 38, 73, 74, 79, 97, 103, 104
        degradation of amorphous silicon modules, 36
        efficiency, 34-38, 41
        high efficiency, 34, 41
        location, 11, 104, 128
        orientation, 12, 75, 104, 105, 128
        seasons, 11
        shadows, 8, 9, 16, 36, 104, 147, 168, 169, 170, 176
        temperature effect, 9, 15, 17, 36, 38, 41, 46, 138, 161, 162, 175
        tracking, 13, 14, 39, 104, 128
    prices, 5
    reliability and warranty, 16, 17, 40, 54, 73, 78, 97, 98, 104
    terminology, 4
    testing, standard test conditions, 15, 118, 149, 161
    types of solar modules, 31
        back contact solar cells and modules, 34
        concentrator modules, 38
        crystalline silicon modules, 32
            polycrystalline, 32, 33, 34
            single crystal, 32, 34, 35, 38
        multi-junctions, 38

INDEX

        thin-film modules, 35
            amorphous silicon (a-Si) modules, 35-38
            cadmium telluride (CdTe), 35
            copper indium selenide/sulfide (CIGS), 35
    why solar, 3
    wiring, source circuits, series, parallel, 20, 63, 176
sources for solar equipment, 115-122
    books, 120
    finding local sources, 115
    installers, 115, 122
    Internet, 115
        coolerplanet.com, 121
        energy.sourceguides.com, 117
        google.com, 115, 118
        gosolarcalifornia.org, 118
        nabcep.org, 121
        seia.org, 122
    kits, 54
    neighbors, 122
    solar exhibits and fairs, 121
    solar magazines, 119
    suppliers, 115, 116
water pumping using solar, 27, 79-98
    choosing a water pump, 97
    control box, 94
    discharge curves, 92
    efficiency, 94
    foot valve, 86, 87
    maximum power point tracking (MPPT), 94
    quality testing, 85
    terminology, 80
    types of pumps, 85
        brushless, 92
        centrifugal pumps, 91
        flexible impeller pumps, 88
        helical rotor pumps (progressing cavity pumps), 90
        jack pumps, 29, 89
        jet-action pumps, 87
        other types of pumps, 92
        piston pumps, 88
        submersible pumps, 86

         surface pumps 87
      storage tanks, 27, 79, 80
      sources of water, 27-29, 82
            springs, 83
            surface water, 84
            wells, 27-28, 82
      types of solar water systems, 27, 79
            battery powered, 27, 29, 80
            direct-connected, 27, 79
            livestock watering, 27, 79
water treatment, 94-97
   air (oxygen), 95
   ozone infusion, 95
   reverse osmosis, 96
   ultraviolet (UV) light, 95

ABOUT THE AUTHOR

Neil Kaminar started his career in solar energy in the early 1970s by helping to design and build a solar house in Atascadero, California. The effort was done in conjunction with CalPoly University. The house used movable insulating panels to control the temperature of water in plastic bags on the roof. The house maintained comfortable interior temperature year round. The sun was used to heat the water in the winter while radiation to the night sky was used to cool the water in the summer.

In the late 1970s, Mr. Kaminar was hired to help design advanced concentrator modules at the Varian Research Center in Palo Alto, California. The concentrators used the advanced gallium-arsenide solar cells developed at Varian. Mr. Kaminar's technical innovations are still the basis for all point-focus concentrators. Mr. Kaminar also helped develop advanced III-V solar cells for space applications, including the first multi-junction cells.

In the late 1980s, Mr. Kaminar formed a partnership company, PVI, for manufacturing concentrators. The company designed, manufactured, and installed linear-focus concentrators. Mr. Kaminar designed the concentrator as well as designed and built much of the manufacturing equipment. Mr. Kaminar held the General Contractor License for the company. He designed and oversaw the installations. In the late 1990s the company was reorganized under the name EcoEnergies and started selling and installing flat-plate modules and systems.

In 2000, Mr. Kaminar joined SunPower Corporation. Mr. Kaminar was part of the team that developed the high-efficiency modules at SunPower. The company had a successful IPO in 2005.

Since 2006, Mr. Kaminar has worked as a consultant to various companies worldwide. His motivation for writing *Solar Basics* and *Solar Design* is to give back to the world some of the knowledge gained in his 40-year career in Solar Energy.

Mr. Kaminar has a Bachelor of Science degree in Mechanical Engineering from Cal Poly University.